Silk, Joseph
Cosmic enigmas.

COSMIC ENIGMAS

Masters of Modern Physics

Published Volumes

The Road from Los Alamos by Hans A. Bethe
The Charm of Physics by Sheldon L. Glashow
Citizen Scientist by Frank von Hippel
Visit to a Small Universe by Virginia Trimble
Nuclear Reactions: Science and Trans-Science by Alvin M. Weinberg
In the Shadow of the Bomb: Physics and Arms Control
by Sydney D. Drell
The Eye of Heaven: Ptolemy, Copernicus, and Kepler
by Owen Gingerich
Particles and Policy by Wolfgang K.H. Panofsky
At Home in the Universe by John A. Wheeler
Cosmic Enigmas by Joseph Silk

COSMIC ENIGMAS

JOSEPH SILK

The American Institute of Physics

In recognition of the importance of preserving what has been written, it is a policy of the American Institute of Physics to have books of enduring value published in the United States printed on acid-free paper, and we exert our best efforts to that end.

Printed in the United States of America.

AIP Press
American Institute of Physics
500 Sunnyside Boulevard
Woodbury, NY 11797-2999

Library of Congress Cataloging-in-Publication Data

Silk, Joseph
Cosmic enigmas / Joseph Silk.
p. cm.—(Masters of Modern Physics; v. 10)
Includes bibliographical references and index.
ISBN 1-56396-061-3
1. Cosmology. 2. Astrophysics.
I. Title. II. Series.
QB981.S553 1994
93-40693
523.1—dc20
CIP

This book is volume ten of the Masters of Modern Physics series.

Contents

NEAR AND FAR

LARGE-SCALE STRUCTURE

DARK AND LIGHT

LIGHT IN THE DARK

About the Series

Masters of Modern Physics introduces the work and thought of some of the most celebrated physicists of our day. These collected essays offer a panoramic tour of the way science works, how it affects our lives, and what it means to those who practice it. Authors report from the horizons of modern research, provide engaging sketches of friends and colleagues, and reflect on the social, economic, and political consequences of the scientific and technical enterprise.

Authors have been selected for their contributions to science and for their keen ability to communicate to the general reader—often with wit, frequently in fine literary style. All have been honored by their peers and most have been prominent in shaping debates in science, technology, and public policy. Some have achieved distinction in social and cultural spheres outside the laboratory.

Many essays are drawn from popular and scientific magazines, newspapers, and journals. Still others—written for the series or drawn from notes for other occasions—appear for the first time. Authors have provided introductions and, where appropriate, annotations. Once selected for inclusion, the essays are carefully edited and updated so that each volume emerges as a finely shaped work.

Masters of Modern Physics is edited by Robert N. Ubell and overseen by an advisory panel of distinguished physicists. Sponsored by the American Institute of Physics, a consortium of major physics societies, the series serves as an authoritative survey of the people and ideas that have shaped twentieth-century science and society.

Preface

Archaeology is the science that probes the past few thousand years. Geology takes us back two or even three billion years. Beyond that is the realm of cosmology: the study of the beginnings of the Universe around us. Thanks to large telescopes on the ground and in space, combined with modern detectors that sample the elusive and sparse photons from remote galaxies, as well as the burgeoning of astrophysics as an independent branch of theoretical physics, cosmology has become a flourishing branch of science.

I began my personal odyssey into the cosmos when as a student I attended some lectures that, while nominally on general relativity, somehow managed to pay a disproportionate amount of attention to an obscure German physicist by the name of Ernst Mach. It was Mach's musings about the meaning of inertia that inspired Albert Einstein to incorporate the universality (or in physicists' jargon, the covariance) of physical laws as a fundamental tenet of general relativity. Einstein elevated the rather rambling thoughts sketched out by Mach into a Principle, promptly dubbed Mach's Principle, that he was determined, unsuccessfully as it turned out, to incorporate into his new theory of gravitation. The inertia of matter, according to this Principle, is determined by the existence and distribution of the most remote matter in the Universe. Any motion measured for the earth can be interpreted as relative to the distant stars.

To many scientists, this seemed, at best, pseudo-science. Then along came the Big Bang theory, culminating in the discovery in 1965 of the cosmic microwave background radiation. This whisper of creation is the greatly faded relic of the primordial fireball that heralded the beginning of the Universe. Not until 25 years had elapsed did the implications of this pivotal discovery reach fruition. The realization came gradually, as the spectrum of the cosmic background was confirmed, with ever-increasing precision, to be indistinguishable from that of a perfect blackbody. Simultaneously, the isotropy of the radiation was studied, and 1992 saw the first definitive announcement of angular fluctuations in the cosmic signal, that were confirmed the following year. Finally, one had a direct connection to the cosmos, and to creation. Only in the first months of the Big Bang would conditions have sufficiently approximated those of an ideal furnace to have generated the blackbody radiation. Only in the first infinitesimal fraction of a second could the fluctuations, observed on scales of tens of degrees, have arisen. Similar fluctuations, on smaller scales, seeded all of the structure in the Universe, including our own Milky Way galaxy.

This cosmic connection defines our place in the Universe, in more ways than one. The uniformity of the radiation in the sky, to better than a thousandth of a percent, tells us that our location can be nowhere special. We are neither at the center nor at the edge of the Universe, but at a perfectly random location in a perfectly isotropic Universe. Yet there is a smooth variation across the sky, reflecting the slight deviation in temperature due to the motion of the earth relative to the cosmic radiation. We are moving towards a direction in Aquarius, where the temperature is consequently slightly higher (by about two-tenths of a percent) than in the opposite direction. Our speed is 370 kilometers per second. This refers to the motion of the earth relative to the most remote depths of the Universe, incomparably further than Mach's "distant stars." The cosmic background is anchored, thanks to gravity, to the matter in the Universe.

Of course, we attribute our motion in part to the pull of gravity by the sun, in part to the Milky Way, in part to the Local Group, and in part to the Virgo supercluster of galaxies. Indeed, even the Virgo supercluster is not at rest but is rushing towards an even larger agglomeration of galaxies that has been dubbed the Great Attractor. Yet motion is inevitably relative, and the final arbiter of the cosmic rest frame is the primordial fireball radiation. In the beginning, apart from the odd speck of matter, this was the Universe. The dreams of Mach and especially of Einstein, little more than shots in the dark, have matured into the Big Bang.

These essays represent my own meanderings around cosmic themes. My topics span the beginning of time until its end, and encompass the enigma of the evolution of large-scale structure culminating in the formation of the galaxies. I have taken these writings from pieces written over the year on various excuses, mostly commissioned to highlight a new look or a new discovery in cosmology. Some have been rewritten, to capture a modern perspective, others remain as written, to encapsulate cosmic thoughts of a decade ago, and some are recent musings. I have kept the original articles almost intact, only using hindsight to correct the few errors of fact. The first two articles are anthologies taken from various book reviews. I have occasionally added brief updates to make the more dated articles timely and relevant. I hope that the mixture is an interesting one.

IN THE BEGINNING

Cosmologists and Their Myths

Mankind has found a niche in the Universe as it expands from an origin shrouded in mystery to an unknown future. The development of human awareness of the Universe evolved from the geocentric cosmology of the ancient world via the heliocentric cosmology of the Renaissance and the egocentric cosmology of the nineteenth century, to the ultimate destination in the Big-Bang theory of the expanding Universe.

The cosmologists of antiquity were brave and persistent souls to tackle the cosmos. The epicycles of Hipparchus and Ptolemy provided a 2000 year respite from the realities of observational astronomy. Full advantage was taken by Thomas Aquinas, who incorporated Aristotle's crystal sphere cosmology into an apparently irrefutable argument for the existence of God. With Copernicus, the heliocentric hypothesis re-emerged and, against all odds, survived, while Tycho Brahe established the first accurate and systematic records of planetary data. Brahe punctured the immutable crystal spheres with his studies of comets and of the famous supernova of 1572. His assistant, Johannes Kepler, derived the laws of planetary motion after a painful decade of sifting endless mathematical permutations of the planetary data, and despite unflinching belief in the celestial harmony exhibited by the five regular solids. Next, Galileo's observations with a 4 centimeter aperture telescope caused an immense stir that led to an unavoidable conflict with the geocentric dogma upheld by theologians. Galileo saw sunspots, the moons of Jupiter and the phases of Venus, all of which established that the heliocentric system was a reality, and could no longer be thought of as a computationally convenient hypothesis. It remained for Isaac Newton to express the universal law of gravitation in a way that gave a theoretical understanding of Kepler's Laws.

The stage was set for modern astronomy. One of the great pioneers was William Herschel, who counted stars and mapped our Milky Way galaxy. Not, however, until the twentieth century was the Sun finally displaced from the centre of the Universe. Harlow Shapley resoundingly overthrew the egocentric view of the place of mankind in the Universe by using the newly calibrated period-luminosity relation for Cepheid variable stars to establish distances to remote globular star clusters. These were found to surround the centre of our galaxy some 30,000 light years from the Sun. Much happened subsequently: the discoveries of the expansion of the Universe and of the cosmic microwave background radiation are two of the highlights that have now led cosmologists almost unanimously to adopt the Big-Bang cosmological model.

In many respects, the Big Bang is to modern cosmology what mythology was to the ancients. To believe that we understand the very early Universe, the first microseconds of cosmic time, requires immense faith in the physicist's search for the ultimate union of the fundamental forces of nature, because direct evidence is completely lacking. Yet to the physicist, the vast energies attained in the immensely compressed primeval fireball that was once the Universe offer a unique testing ground for the latest theories of elementary particles. Unification of the fundamental forces of nature is the goal, although gravity remains the poor partner. Supersymmetry is conjured up, doubling the number of elementary particles with the stroke of a pen, and thereby providing a virtually inexhaustible zoo of dark matter candidates. *SUSY GUTS* is not a radical feminist, but a theory that is designed to unify the fundamental forces of nature. Exotic names, like photinos, textures, strings or superstrings, are reeled off as though they were the most natural state of matter, which perhaps they once were. Provided we accept theories of matter and gravity that are certainly correct at our present epoch and in our environment, then we are inevitably led by the Big Bang back to a singular state near the origin of time.

Of course we are still awaiting the ultimate Theory of Everything, which will explain the most important missing link in the puzzle, namely how it all began. Many physicists around the world are pursuing this ultimate goal that represents the union of gravity and quantum mechanics. However, we are not there yet, and opinions differ widely as to how far away we may be: according to Stephen Hawking, the end of physics is in sight, yet Sheldon Glashow argues that superstrings are only a mirage.

Cosmology, perhaps more than most disciplines, has attracted individualists who already had outstanding achievements in related fields such as astronomy, physics, or even theology. Some appeared like meteors, blazing a path through the popular press, but their contributions invariably faded into obscurity. And even many of the great cosmologists of the past spent the most productive years of their careers searching in vain for their Holy Grail, the elusive Theory of Everything. Edward Milne and logarithmic time, James Jeans the great popularizer, and Fred Hoyle of steady state fame are in the first category. Arthur Eddington and Albert Einstein are in the second.

After hearing Eddington propound his alleged derivation of the fine structure constant from fundamental theory, one eminent physicist in the audience wondered whether all physicists go off on crazy tangents when they grow old. His colleague explained: "You don't have to be scared. A genius like Eddington may perhaps go nuts, but a fellow like you just gets dumber and dumber".

The hero of modern cosmology is undoubtedly Monsignor Georges Lemaître. This Belgian cleric and mathematician was the greatest pioneer of modern cosmology. Born in 1894, he had an early interest in relativity theory, spending the academic year 1923/24 with Eddington in Cambridge and then travelling widely in the U.S.A. He produced his first fundamental paper propounding the expanding universe theory in 1927, unaware of parallel mathematical results by Alexandre Friedmann in Petrograd five years previously. However Lemaître, unlike Friedmann, made the crucial and far-sighted connection with the galaxy redshifts of Vesto Slipher, despite positive discouragement from the contemporaneous giants of the field, Einstein and Willem de Sitter. He spent his later years exploring the consequences of an early dense phase of the Universe: many of his ideas, though now largely superceded, have echoes in the modern concepts of quantum gravity accounting for the beginning of the Universe, relics of the early phase producing the cosmic blackbody radiation, and light element synthesis in the first few minutes. He was far ahead of his time: even the greatest of contemporaneous observers, Edwin Hubble, discoverer in 1929 of the law relating galaxy redshift to distance that underpinned the new cosmology, was never able to fully accept the physical reality of the expansion of the Universe. Lemaître truly was the first physical cosmologist. One can only imagine his excitement when on his deathbed in 1965, he heard of the ultimate footprint that he had sought for the Big Bang, definitive evidence fossilized as a dim relic of the primeval fireball, the cosmic microwave background radiation.

One of the greatest pioneers in astrophysics, far ahead of his time, was Fritz Zwicky. Despite valuable contributions to the early development of jet propulsion, his application for U.S. citizenship was rejected in 1949 when he was forced to resign as research director of the Aerospace General Corporation. Thereafter he plunged full tilt into astronomy, and his galaxy catalogs laid a solid foundation for modern studies of the large-scale structure of the Universe. Zwicky taught theoretical physics at Cal-Tech from 1927 to 1942, when he made fundamental and farseeing conjectures about the existence of neutron stars and the prevalence of dark matter in the Coma cluster of galaxies. Zwicky had unconventional notions about the significance of objects he catalogued, such as blue compact galaxies, but his data have stood the test of time. Partly because of personality conflicts, he remained at odds with the scientific establishment, as do those today who refuse to accept the Big-Bang theory.

One of Zwicky's major themes was that it pays to recognize a paradigm shift. The cosmic microwave blackbody radiation, a perfect Planck function (Nobel Committee take note: this deserves recognition!), provided the death knell to sceptics of the hypothesis of creation in the first instants of time. Proponents of alternatives, such as a plasma universe (whatever that may be) or a reincarnated steady state universe, are in the situation today of myth propagators. Appealing to virtual deities to support the crystal spheres of their incoherently formulated theories, they invariably speak up at scientific meetings to try to counter the bandwagon of the Big Bang. These Jeremiahs wail that we still await the ultimate theory of the beginning of time, that galaxy formation is as yet poorly understood, that we cannot reconcile the smoothness of the cosmic microwave background with the large scale structure of the universe. The weakness, inevitably, is selective vision, culminating in a virtually fraudulent statistical misuse of data to prove that Hubble was wrong, or that the Big Bang never occurred.

The paradigm has shifted since the 1920's: from static to expanding universe. Modern cosmology and the Big-Bang theory are in surprisingly good shape. We cannot yet predict the weather: galaxy formation is a more complex phenomenon, and we are continually refining our theories to keep up with the unremitting flood of new data. There is absolutely no doubt that the Big Bang is here to stay, although our understanding of the first milliseconds of the Universe may well change beyond recognition in future years. There is a lesson to be learnt from studying the great cosmologists of the past, including Georges Lemaître and Edwin Hubble. We should be unceasingly open to radical new ideas, but only when based on sound physical principles or new, statistically significant, observational data. Sadly, the new sceptics fail dismally to match the intellectual brilliance of that ultimate scientific outsider, Fritz Zwicky, who concluded that "after Pythagoras had discovered his famous theorem, the Greeks slaughtered 150 oxen and arranged for a feast. Ever since that happy time, however, whenever anybody proposed something drastically new, the oxen have bellowed".

Yakov Borisovich Zel'dovich was a cosmologist who never strayed far from physics. He was also one of the world's leading physicists whose interests spanned many branches of physics. Indeed, Zel'dovich pioneered exploration of the implications of the microwave background for cosmology. Together with his former student Rashid Sunyaev, he developed the theory of the shadowing of the background by hot gas in galaxy clusters. Zel'dovich argued that one could account for the creation of the Universe from nothing, or more precisely, from almost nothing. The ultimate Theory of Everything is needed to describe the first instants of the Universe, where all known physical laws break down before we can understand what creation truly involves. Perhaps it is spontaneous creation from "anything", as likely a precursor to the observable Universe as "nothing". It is sad that the COBE detection of fluctuations in the cosmic microwave background, a unique probe if not a proof of inflation, was announced less than five years after Zel'dovich's premature death.

Zel'dovich's love for cosmology is reflected by the many pithy and personal anecdotes and comments that are sprinkled throughout his writings. He attributed some to others, e.g. "cosmology is often in error but never in doubt" (Landau), or on the bandwagon effect of cosmological theorists on astronomers to lay undue emphasis on visual impressions of their data: "Turner's paintings have affected England's climate: the numbers of clear evenings having brilliant sunsets over a rolling sea has increased dramatically" (Wilde). But some are his: on baryon asymmetry "almighty God throwing dice for every single proton or antiproton would soon get tired with the astronomical number of particles. He could not make the asymmetry large enough". On highbrow theoreticians, he concedes that "the ultimate topic—the very birth of the universe—remains out of reach of even the highest brow", and on scalar fields "once the genie has been let out of the bottle, he never wants to go back inside".

The ultimate difficulty is that cosmology is a science starved of data and not readily amenable to controlled experiments. We must take each glimpse of the Universe as it comes, no matter how confused or obscure it happens to be. Remote galaxies appear as they were aeons ago, and thereby profoundly alter our normal sense of perspective if, for example, young galaxies were more luminous than the mature nearby galaxies. It is as though we view the Universe through a distorting lens, and we have little sense of the nature of the distortion. And this applies to the galaxies, whose formation was only the most recent episode in the history of the Big Bang. Extrapolating backwards in time from the furiously receding galaxies we see today, we infer that the Universe was once a far more exotic, denser, hotter and hostile place.

Anthropic Musings

Once upon a time there was a flea, who lived on the back of a dog called Bonza. The flea thrived, and grew as big as a pea, for Bonza's fur was the perfect environment, neither too hot nor too cold, too dry or too moist, too dark or too light. Then one day, the dog's mistress purchased a flea collar for her pet. From that day on, the world was a very harsh place for the flea, who no longer had a moment's peace. Sadly, the flea was forced to jump, and landed by a lamp-post, where it soon froze to death.

The moral of this story is that although a canine universe may be eminently well suited for the existence of fleas, fleas are not inevitable. Some dogs have them, some do not. It is equally presumptuous to believe that man's existence is inevitable, as advocated by proponents of what has become known as the strong anthropic principle. The Universe does provide a suitable environment for man to exist, but the flea's precarious existence should caution us that even this weak anthropic argument is more tautology than teleology. Bonza's fur provides a breeding ground for fleas quite independently of whether or not fleas exist.

This current cosmological theme provides a remarkable bridge between the philosophical undertones associated with the origin and fate of the Universe and the human perspective. According to the anthropic principle, the properties of the Universe are determined by our presence as observers. Conditions must be congenial for intelligent life to evolve. This means, for example, that the early Universe could not have been highly irregular, inhomogeneous or chaotic, nor however could it have been perfectly regular, otherwise galaxies and stars would not have formed. The anthropic principle accounts for the various large-number coincidences involving powers of 10^{40}, the ratio of electromagnetic to gravitational coupling constants. These include the mass of a star and the mass of the observable Universe, as well as the coincidence between nuclear, stellar and Hubble expansion time-scales. It even purports to account for the approximate values of the constants of nature: if they differed significantly from observed values, human beings could not have evolved.

The Universe is isotropic and homogeneous and long-lived and matter-dominated, all ingredients more or less necessary for man's existence, because of events that occurred a fraction of a second after the Big Bang. Cosmologists continue to debate why those events happened, and precisely how they proceeded, but their conclusions are not going to account for the origin of life. Nor should the procession of unlikely events that heralded the evolution of life be allowed to get cosmologists off the hook: they still have much explaining to do.

A number of cosmologists in their more literate moments lend a sympathetic ear to this sort of teleological tattle. The complexities of nature offer a glimpse of a grand designer, or rather, of a grand design, with the designer absent, essentially in hiding. Existence of a watch implies that there was a watchmaker: we are urged to take the logical next step, that existence of a flower implies that a cosmic flowermaker must once have plied his craft. From the flower, it is simple to inquire about the origin of a sentient being, and presto, we have proven the existence of a deity.

There is, of course, a flaw in this logical progression, and I wish that all budding grand designers would take due note. We know how to make a watch, hence we correctly infer the existence of a watchmaker. Our cleverest botanists do not know how to create a daisy, and we are still farther from constructing a chimpanzee. Hence we can conclude nothing about their designers: we lack the information necessary to make any inference. A reductionist approach fails utterly to account for a solitary daisy. The laws of physics, applied to the individual molecules, could succeed only with negligible probability. Some additional ingredient is needed. Is this necessarily where the theologians enter?

Do you prefer the proof of the existence of God by St. Thomas Aquinas? Or do you defer to the derivation of the wave function of the Universe by Stephen Hawking and James Hartle? How about the creation myth of the Boshongo, a Bantu tribe? Or does evolution in an 11-dimensional Kaluza-Klein universe more accurately describe the origins of space and time? The unprejudiced observer may find it difficult to choose between these options.

Teleology is the notion that the Universe was shaped not by chance but by a grand design. After Darwinian evolution had destroyed much of traditional teleology, cosmology developed into a physical science during the twentieth century, with the introduction of the tensor calculus helping to drive the philosophers and theologians out of the field, much to its subsequent loss. But the pendulum does swing in science, and we now have Brandon Carter and Frank Tipler making much of the idea that our presence is indispensable to the existence of the observed universe. Even more, our presence requires a unique universe, thereby coupling the largest observable horizons of tens of billions of light-years to our puny solar system a mere light-hour across. Even the infinite horizons that the future holds in store for us are uniquely constrained by our mere act of existence.

Biological application of the anthropic argument in a form that originated with Brandon Carter leads to a relation between the number of improbable steps that have enabled humans to evolve and the future length of time during which the biosphere will continue to evolve. Remarkably, we learn that intelligent life on Earth may only be destined to evolve for another 40,000 years. It is not difficult to think of a mechanism by which life could be extinguished, but how seriously do we take, for example, the inevitability of nuclear winter? Carter's hypothesis relies on the improbability of the evolution of intelligent life. Should our galaxy turn out to be teeming with intelligent extraterrestrials, all bets are off. John Barrow and Frank Tipler have argued, however, that space colonization by such extraterrestrials would have been inevitable and that any such species would have revealed itself over the billions of years available for its self-replicating technology to have developed. Of course, the sceptic may wonder whether a truly intelligent species could have developed a plethora of means to hide all trace of extraterrestrial contact.

A more rational approach is worthy of serious consideration. Unpredictability is a vital element of many non-linear systems. Push the simple pendulum just so far, and it no longer exhibits its predictable swing. The turbulent patterns, whether of cream stirred in coffee or of cumulus cloud formations, are infinitely more complex than their basic elements (coffee, cream or water vapor). The theory of the behaviour of simple dynamical systems provides us with an understanding of how non-linearity can trigger seemingly chaotic patterns. It is when one goes from physical to biological systems that an intriguing question arises. Is biological complexity mechanistic in origin, or do we need to go beyond the laws of physics and chemistry to account for the diversity of a living system? Perhaps the non-linearities that arise when a complex broth of primordial amino acids is suitably processed will induce sufficiently non-random mutations and syntheses to lead to otherwise highly improbable primitive life forms.

The real challenge is yet to come. Is a sentient being no more or less than a highly complex non-linear dynamical system? Certainly, the implausibility arguments advanced by, among others, Frances Crick and Fred Hoyle, which purport to demonstrate the need for an extraterrestrial origin of life, are no longer viable. They must be abandoned in the face of chaotic behaviour and the ensuing unpredictability of dynamical systems. Whether consciousness, which is an extreme form of organization, can arise in this manner cannot be answered, if we do not know the capacity of matter for self-organization.

It is essential in all of these semi-philosophical meanderings to distinguish science from pseudo-science. For example, pseudoscientific overusage of the second law of thermodynamics has taken a life of its own. Metaphors are fine, but applications to society and even biology seem hopelessly flawed. It is just as well that someone has challenged the propagator (who else but Jeremy Rifkin?) of such statements as "if love were antientropic, it would be a force in opposition to becoming, for the entropic flow and becoming go hand in hand," or "the governing principle of a low entropy world view is to minimize energy flow. . . . a low entropy society deemphasizes material consumption," or again, "new genetic technologies, like recombinant DNA, might greatly increase the matter energy flowing through the system, just as the first industrial revolution did with renewables," or to balance this, "the practice of meditation is designed to slow down the wasteful expenditure of energy." Somewhere along the way, "entropy" and "energy" have gotten hopelessly confused, but this is the least of one's worries: after all, as Tony Rothman noted in his book "A Physicist on Madison Avenue," Clausius said their meanings are nearly identical. Nearly, but not quite: that ultimately is the distinction between science and pseudoscience.

Where does this leave us? The uniqueness of our Universe can be attributed to our existence as observers, if we accept the anthropic principle. But have we really advanced our understanding of any aspects of cosmology? In 1925, A.N. Whitehead perceived about cosmology that "there is no parting from your own shadow". Is the Universe merely our shadow? Or perhaps the cosmologists are deceiving themselves and the roles are reversed. The truth may very well be that with only one Universe to explore, we may never resolve this paradox.

The Big-Bang Theory

Cosmology is the study of the large-scale structure of the Universe, and of its evolution and origin. The central hypothesis of modern cosmology is the Big-Bang theory, which postulates that the Universe began in a fiery explosion that occurred some 15 billion years ago. Modern cosmology is a twentieth century phenomenon that grew from Albert Einstein's theory of gravitation (1916) and Edwin Hubble's early exploitation of the Mount Wilson 100 inch telescope in the 1920's. In recent years, the field of cosmology has grown dramatically as ever-more powerful telescopes are brought to bear on distant galaxies in a variety of wavelength bands. Evidence for the Big Bang has accumulated, and the theory has been refined to the extent of accounting for the first seconds of the Universe. Our knowledge of large-scale structure has developed to the point that we finally have a fair sample of cosmic structure out to a depth of 100 megaparsecs or more. Our theoretical understanding of the origin of structure is a central element of the expanding universe theory that has culminated in the 1992 detection of fluctuations in the residual fireball radiation from the Big Bang. Some issues remain unresolved.

Historical Developments

In 1915, Einstein announced a new theory of gravitation, general relativity, that identified gravity with the curvature of space-time, the four-di-

---

This has also appeared as "Big-Bang Theory," Joseph Silk, *Encyclopedia Americana*, (1993).

mensional manifold that consists of the three space dimensions combined with time. Any event can be described in terms of its location and path in space-time, and in particular, the light from distant galaxies follows the shortest path, a geodesic. One looks back in time along geodesics: galaxies are time machines and the light from the most distant galaxies has been travelling through space-time since before the earth formed, 4.6 billion years ago. The distance to these galaxies is measured in light years, the most distant galaxies being at a distance of 10 billion light years, or a look-back time of 10 billion light years ago.

Einstein's theory of general relativity received dramatic confirmation in 1919, when the deflection of light from distant stars by the sun was measured during a total eclipse, and the cosmological implications of his theory received intensive study. Einstein himself in 1917 had already developed a model of cosmology in which he postulated a cosmic repulsion force that was characterized by a new term, the cosmological constant, in his gravitational field equations. The role of the cosmic repulsion was to balance gravity and yield a static model for the Universe. However, the addition of a cosmological constant was soon shown to be unnecessary, there being a cosmological solution to Einstein's equations that Einstein himself had overlooked.

In 1922, the Soviet meteorologist and mathematician Alexandre Friedmann, and in 1927, the Belgian cleric Abbé Georges Lemaître independently demonstrated that the Universe could be in a state of large-scale expansion (or contraction). To avoid collapse, expansion of the Universe balanced gravitational attraction. The expansion could continue either forever, if the matter content was insufficient to decelerate the expansion sufficiently, or could eventually reverse into a future phase of contraction. A principle implication was that the matter content of the Universe implied that space was not necessarily Euclidean or analogous to the flatness of a plane in a two-dimensional analogy, but could be curved, as the surface of a sphere (positive curvature) or a hyperboloid (negative curvature). Since the surface of a sphere is finite and closed whereas that of a hyperboloid is infinite and open, we infer that a universe with high matter density should be a positively curved, closed, finite universe that eventually collapses, whereas one with low matter density is a negatively curved, open, finite universe that expands forever.

Hubble's discovery of a linear relation between distance to a remote galaxy and its redshift in 1929 provided dramatic evidence that the Universe indeed was expanding. The redshift is due to the Doppler shift of light from a receding galaxy, and provides a natural explanation for Hubble's law if it is space itself that is expanding as had been predicted in particular by Lemaître. While Hubble's original data only sampled nearby galaxies to within 10 megaparsecs, the linear expansion law has subsequently been confirmed to beyond 100 megaparsecs.

In 1931, Einstein and the Dutch astronomer Willem de Sitter proposed that the Universe could be at critical density and spatially flat, exactly intermediate between these two Friedmann-Lemaître models. Such an Einstein-de Sitter universe is infinite, has the geometry of a plane, with Euclidean (or flat space) geometry, and also expands indefinitely. Under the assumption that the Universe is approximately homogeneous and isotropic on large scales, known as the cosmological principle, H.P. Robertson and A.G. Walker found in 1936 that there are only three possibilities for the geometry of space-time.

These three choices reduce to the uses of positive, negative and infinite curvature radius for three-dimensional space, corresponding to a supercritical, subcritical and critical mean density of matter in the Universe. The critical density, that of the Einstein-de Sitter cosmology, can be simply expressed as $3H_o2/8\pi G$, where H_o is Hubble's constant of proportionality between recession velocity and distance, and G is Newton's constant of gravity. Cosmologists have not yet succeeded in measuring the observed density with sufficient precision to restrict this choice of possible models.

The cosmological principle itself has received impressive verification over the past three decades, through studies pioneered by Hubble in the 1930's of deep galaxy counts that demonstrate homogeneity, and mapping of the cosmic microwave background radiation discovered in 1965 that reveals a remarkable degree of isotropy. The Shane-Wirtanen deep galaxy survey analyzed in the 1970's by Peebles and collaborators contained about one million galaxies. The most recent survey (1990) uses sophisticated computerized data reduction techniques developed by George Efstathiou and collaborators, and contains some four million galaxies. The number density of galaxies is found to increase with increasing survey depth as would be expected if, on the average, the galaxy density is spatially uniform. The large-scale isotropy of the Universe is now measured to a precision of about 0.001 percent via the microwave background.

The Big Bang

The Friedmann-Lemaître models are described at epoch *t* by a universal scale factor *a(t)* that represents the variation with time of the distance between any pair of remote objects in the Universe. The observed density of the Universe is at least 10 percent of the critical value. We consequently may infer that when the Universe was less than about one-tenth of its present size, the competing models (open, closed) were indistinguishable from the Einstein-de Sitter model.

The Hubble constant H is really not constant, since H equals the ratio of velocity, or rate of change of distance to distance. H is the expansion rate of the Universe, and is inversely proportional to t. At an early epoch, t was small and the expansion rate was enormous: hence the early Universe expanded explosively, justifying its description as the Big Bang.

George Gamow and his collaborators, commencing in the late 1940's, realized that the dense state of matter at an early epoch that was implied by the Big-Bang theory could have observable implications. The Universe might have been as hot and dense as the interior of a star. This means that nuclear reactions would have occurred. The primordial constituents of the Universe were first shown by Chushiro Hayashi in 1950 to consist of electrons, protons, neutrons, neutrinos and photons. Nuclear equilibrium means that there was about one neutron for every ten protons. Gamow's key contribution was to demonstrate that the neutrons combined with protons to form helium. Since a helium nucleus contains two protons and two neutrons, the Big-Bang model predicted a universal abundance of helium, the second most abundant element (after hydrogen) in the Universe of about 25 percent by mass.

Gamow's original goal of synthesizing the heavier elements failed, because there are no stable elements in the periodic table of masses 5 and 8 that constitute the sequence for the natural building blocks of hydrogen (mass 1) and helium (mass 4). Trace amounts of deuterium (mass 2) and lithium (mass 7) are produced in the first minutes of the Big Bang. After a few minutes, the density and temperature have fallen and nuclear reactions ceased, leaving hydrogen and the synthesized light elements as primordial matter to form the first stars. Elements such as carbon, oxygen and iron are synthesized in stellar interiors, where the long time-scale allows synthesis of carbon by three-body nuclear interactions (three helium nuclei forming a carbon nucleus) to bridge the gap between light and heavy element nucleosynthesis.

Light element synthesis in the Big Bang required that the temperature of the Universe during the first minutes be some hundreds of millions of degrees Kelvin. This led Gamow to predict that such a primeval fireball, when expanded to fill the observed Universe today, would still be present but at greatly reduced intensity, corresponding to a highly redshifted wavelength. He estimated that the present temperature would have fallen to several degrees Kelvin. Unaware of Gamow's prediction, Robert Dicke and his collaborators participated in a search at Princeton University for the relic fireball radiation in the early 1960's. Dicke conjectured that the radiation must have a blackbody spectrum and would be observable at microwave and radio wavelengths. In fact, the cosmic microwave background was discovered in 1965 by radio astronomers Arno Penzias and Robert Wilson at Bell Laboratories, in New Jersey. Dicke confirmed the discovery, and a new era began in cosmology.

Evidence For The Big Bang

There are three principle pillars of support for the Big-Bang theory. The expansion of the Universe according to Hubble's law provides direct evidence for an explosive beginning. The predictions of light element synthesis in the first minutes of the Big Bang have been verified by observations of the light element abundances. The helium fraction relative to hydrogen, when corrected for production of small amounts of helium by stellar nucleosynthesis, is remarkably constant from galaxy to galaxy, testifying to its cosmological origin. Trace amounts of deuterium, which is only destroyed in stars and not produced, are observed in the interstellar medium that are concordant with the Big-Bang prediction. Lithium is yet another element whose abundance correlates not with other metals but is observed to plateau in the oldest stars in our galaxy, characteristic of a pregalactic origin. The abundance predictions are quite robust: for example, the number density of protons is constrained to be a few percent of the critical density.

Any deviations from Friedmann cosmology from the first second onward must consequently be small. This led to the prediction that at most four types of neutrinos can exist, since these light particles contribute to the energy density and thereby the expansion rate of the Universe. In 1990, the LEP experiment at CERN showed that there are precisely three types of neutrinos by monitoring the decays of unstable heavy particles, Z bosons, that link electromagnetic and weak nuclear reactions. This inspires confidence in the Big-Bang theory to an epoch as early as 10^{-10} second, when the Z bosons last existed in large numbers. At this epoch, the temperature was 10^{15} degrees Kelvin and equivalent to an energy of about 100 GeV, corresponding to the mass of a Z boson.

The most definitive support for the Big Bang has come from the cosmic microwave background. The Universe has now cooled to a temperature of about 3 degrees Kelvin, and the radiation peaks in the microwave spectral region, at a wavelength of about 1 millimeter. Experiments in 1991 measured the spectrum of the cosmic microwave background to unprecedented precision. It was found to deviate near the blackbody peak by no more than three-hundredths of a percent from a blackbody of temperature 2.726 Kelvin. The cosmic microwave background was found to also be highly isotropic. A dipole (or 180 degree) variation in temperature due to the motion of the Earth at a level of about 1 part in 1000 was discovered in 1973. However only in 1992 were angular fluctuations reported on angular scales between 10 and 90 degrees, amounting to about 30 millionths of a degree Kelvin, or 1 part in 10^5.

The blackbody spectrum is most simply understood in terms of an origin of the radiation at an epoch when the Universe was hot and dense enough to be an efficient furnace: this happened within the first months of the Big Bang. The extreme degree of isotropy further points to an origin of the microwave background in the distant universe: any more local origin would almost inevitably reveal structure in the form of appreciable angular variations in temperature.

Inflation

The observed isotropy of the microwave background is not simply understood in terms of the original Friedmann-Lemaître theory of the Big Bang. Light from one side of the sky has not had time to traverse the opposite side of the sky, yet the temperature of the microwave background in these two directions is identical to within measurement error.

A physical explanation for this uniformity was provided in 1981, when Alan Guth discovered inflationary cosmology. As the Universe expanded below the grand unification (GUT) scale at a temperature of about 10^{16} GeV (or 10^{29} degrees Kelvin), when unification of the strong, weak and electromagnetic forces occurred, the pressure of matter suddenly decreased. In fact, the pressure became negative (as in a crystal lattice) for a brief phase, and triggered a rapid increase in the expansion rate. Normally, positive pressure acts as energy density and its gravity helps decelerate the expansion of the Universe. The expansion rate became exponentially rapid between about 10^{-35} and 10^{-33} second during an epoch of inflation. The pressure soon reverted to its normal value and the Friedmann expansion subsequently resumed. Inflation meant that large scales, far in excess of the present light travel distance, were once in causal contact. One can therefore understand the observed uniformity of the Universe.

Inflation may not be the only explanation: an alternative hypothesis simply asserts that the early Universe was uniform at the epoch of the Big Bang itself. However, inflation has at least one other prediction. Any initial curvature of space is smoothed out by the exponential expansion, and inflation predicts that the Universe should presently be at critical density, corresponding to a Euclidian geometry. Since the observed density of matter in the form of baryons and luminous stars is inferred from the primordial nucleosynthesis to be at most 10 percent of the critical value, and more likely to lie between 3 and 6 percent, one concludes that at least 90 percent of the mass of the Universe must be non-baryonic, if inflation occurred.

Dark Matter

The prime candidate for non-baryonic dark matter is a weakly interacting elementary particle, produced in great numbers in the early Universe. If the interactions of such particles are sufficiently weak, not all will annihilate with one another as the Universe expands and cools, leaving a sufficient number behind to account for the dark matter. Certain theories, such as supersymmetry, predict the existence of these weakly interacting particles, and accelerator experiments are able to search for them, so far with no success.

Astronomical experiments are also underway to detect weakly interacting non-baryonic particles. The Earth moves through the dark halo as the sun orbits the galaxy. If our dark halo consists of such particles, their flux on the earth is known. Cryogenic laboratory experiments are being designed to search for the nuclear recoils and ionization that is induced in sensitive detectors such as pure silicon.

An alternative possibility is that the dark matter which is known to constitute the major component of galaxy halos is baryonic. Possible forms of baryonic dark matter include stellar relics such as white dwarfs, neutron stars or black holes, as well as objects of substellar mass, known as brown dwarfs. One would still expect galaxies to be swimming in a more uniform sea of non-baryonic dark matter if inflation is correct.

Origin Of Structure

Within the microwave background, we detect about 10^9 photons for every baryon, and at a sufficiently early epoch, photons dominate the mass density of the Universe. One finds that before an epoch of about 10^3 year or a temperature of 10^5 Kelvin, the energy density of the Universe was dominated by radiation.

A legacy of inflation was the generation of density fluctuations with a nearly scale-invariant distribution, there being no preferred scale. One cannot yet calculate from the theory of inflationary cosmology the amplitude of the primordial density fluctuations, but their scale invariance is predicted by the simplest models of inflation. The process of gravitational instability, by which the self-gravity of density fluctuations helps accrete nearby matter and thereby enhances the amplitude of the fluctuations, is responsible for converting primordial fluctuations into the large-scale structure of the Universe. However, there is a long delay before growth occurs. Radiation dominance inhibits growth, self-gravity only aiding fluctuation growth once ordinary matter is the dominant component.

One consequence is that a natural scale, that of the horizon or maximum size of a causally connected region, is imprinted onto the primordial fluctuation spectrum in the so-called cold dark matter models, those in which the dark matter consists of weakly interacting particles with negligible velocity dispersion that aggregate freely into the growing density fluctuations on galactic scales. These fluctuations eventually develop into large clouds of matter that are the precursors of galaxies and clusters of galaxies, after a few billion years have elapsed. We can see the microwave background back to a temperature of about 3000 Kelvin or to an epoch of 300,000 years. The hydrogen was ionized, in plasma form, prior to this time. During the plasma phase, the photons underwent frequent scatterings. Once the hydrogen combines into atoms, the radiation propagates freely. The primordial density fluctuations therefore leave their imprint as fluctuations in the cosmic microwave background.

Unresolved Issues

The COBE discovery of fluctuations on angular scales between 10 and 90 degrees may be a consequence of an early inflationary period, the only causal way that is known to induce fluctuations on such large scales. Such large scales, however, greatly exceed the scales of the precursors of present day structure, expected to be on an angular scale of 1 degree or less. The inflationary theory predicts, however, that these large angular scale fluctuations would have formed from quantum fluctuations amplified by inflation, the same process that generated the precursors of the observed large-scale cosmic structure. Discovery of pregalactic fluctuations will be a crucial step in confirming the gravitational instability theory of structure formation.

The detection of fluctuations at large angular scales fixes the normalization, or amplitude, of the primordial density fluctuations. This poses a challenge for theories of structure formation. With massive, weakly interacting relic particles as dark matter, one finds that the dark matter is generically cold at the epoch of structure formation, and all of the dark matter is destined to collapse into galaxy halos and clusters.

In fact, one does not observe a critical density of matter in the form of any observed structure on galaxy or galaxy cluster scales. To make the theory work if the Universe is at critical density, theorists invoked *biasing:* the assertion that dark matter fluctuations are weaker than those seen in galaxy counts. Most dark matter would then remain more uniformly distributed than the matter in halos and clusters of galaxies. There is still a lack of power on intermediate scales, between 10 and 100 megaparsecs, as is apparent from the occurrence of superclusters, voids, and especially the bulk flows whereby clusters of galaxies have peculiar motions, relative to the Hubble expansion of several hundred kilometers per second.

Nevertheless, cold dark matter has led to a reasonably successful theory of galaxy formation, provided that the rms fluctuations are about half the amplitude measured by COBE. In contrast, there is the possibility of hot dark matter. Here the dark matter, perhaps most naturally, consists of light (10–100 electron volt) neutrinos that were effectively massless and moving at light speed, hence hot, at the temperature of 10 electron volts when structure could first have developed. Consequently, galaxy formation is strongly suppressed in a hot dark matter model, and cold dark matter has generally superceded hot dark matter as a viable cosmology.

However, the new normalization tells us that on large scales, dark matter is unbiased, mass tracing light. This has created somewhat of a quandary on smaller scales, where galaxies form. Several solutions have been proposed, that range from invoking an admixture of cold and hot dark matter, and appealing to the uncertain physics of galaxy formation. The most radical proposal is that the large-scale fluctuations in the microwave background are due to very long wavelength gravitational waves left over from inflation, redshifted to the scale of the present horizon. Were this the explanation, and inflation predicts the existence of such waves although with highly uncertain amplitude, the primordial density fluctuations would not yet have been measured by COBE. One would therefore have to await the results of searches on smaller angular scales where gravity waves are unimportant in order to definitively measure the amplitude of the primordial density fluctuations that seeded structure formation.

The Very Early Universe

WITH JOHN D. BARROW

The cosmological principle is the powerful concept that the Universe is homogeneous and isotropic. In other words, the large-scale features of the Universe would appear the same to an observer in any galaxy no matter in which direction he looked. Much observational and experimental work supports the cosmological principle, which is deeply rooted in physics and natural philosophy. How did the Universe acquire its large-scale uniform structure? Either it has pretty much always been that way or it was highly irregular and chaotic right after the Big Bang and has evolved into its present form because of certain smoothing and heating mechanisms. According to the latter possibility, called chaotic cosmology, the smoothing and heating mechanisms would give rise to the current regular Universe regardless of the extent of the initial irregularity. Therefore chaotic cosmological theories eliminate the vexing problem of having to know the initial conditions of the Universe.

Attractive as the elimination of the problem is, such theories may have a fatal drawback. We believe the proposed smoothing and heating mechanisms would irreversibly generate more thermal energy than seems to exist in the Universe today. We think the Universe as a result had only an infinitesimal degree of irregularity at the time of its creation. What happened at the precise moment of creation is not yet known because unfamiliar physical principles unique to the immense densities and temperatures of that moment mask the initial structure of the Universe. Matter behaves in a way that gives little inkling of what the Universe was like in the first 10^{-35} second of its existence.

This originally appeared as "The Structure of the Early Universe," John Barrow and Joseph Silk, *Scientific American*, 98–108, April 1980.

The cosmological principle gains observational support from the fact that the Universe is undergoing an expansion in which every galaxy cluster is rushing away from every other. In 1923 Edwin P. Hubble discovered that the rate of expansion increases with distance from the observer. He detected the recessional motion of the distant galaxies through measurements of their optical spectra. The wavelength at which electromagnetic radiation from a distant object reaches the earth is increased by the velocity of recession of the object with respect to the observer. This is the well-known red shift, so named because if the radiation from the receding object is in the visible region of the spectrum, it is made redder.

The amount of the red shift is given as a number corresponding to the fractional increase in the wavelength of the received radiation. For example, galaxies 20 million light-years away (among the closest to our own galaxy) have a red shift of .001 and galaxies 10 billion light-years away (among the most distant) have a red shift of .75. By measuring the red shift Hubble was able to calculate the recessional velocities of distant galaxies and thus the expansion rate of the Universe.

The nature of the expansion can be understood by a traditional visual metaphor: likening the Universe to a spherical balloon with dots painted on its surface, each dot representing a galaxy. As the balloon is inflated the distance between any two dots (as measured on the surface of the balloon) increases at a rate proportional to the distance between them. An observer at any dot would see all the other dots receding from him uniformly in all directions; no observer would occupy a privileged position. To put it another way, the expansion has no center.

It is not known whether the expansion of the Universe will continue forever or whether someday the galaxies will stop receding from one another, start moving in the opposite direction and eventually fall together. Either possibility is consistent with prevailing cosmological theory, which maintains that the Universe began with an explosion from a super-dense state. The type of expansion is determined by the space-time geometry of the Universe. Infinite expansion means an "open" universe; finite expansion followed by collapse means a "closed" universe. The critical intermediate case, advocated by Albert Einstein and Willem de Sitter in 1932, is where the Universe has the minimum energy needed to overcome the decelerating influence of gravity and expand forever to infinity. Whether the Universe is open or closed is difficult to determine because the expansion energy is near the critical value. This is the subject of continuing investigation in astronomy.

The Universe is isotropic and homogeneous not only in its rate of expansion but also in its distribution of constituent objects. Hubble counted the number of distant galaxies in separate quadrants of the sky and in different volumes of space. He found that the larger the volume is, the more galaxies it contains. Moreover, the distribution of galaxies with direction varied hardly at all. More recent surveys have probed the uniformity of the distribution of galaxies in the Universe to much greater distances. For example, when distant regions with a radius of a gigaparsec, or 3×10^9 light-years, are compared, their populations of radio-emitting objects (galaxies and quasars) are found to be equal to within 1 percent.

The most compelling evidence of isotropy comes from the background of microwaves, or radio waves with wavelengths in the millimeter range, that seem to flood the entire Universe. The radiation was discovered in 1965 by Arno A. Penzias and Robert W. Wilson of Bell Laboratories. Sky maps of microwave radiation based on measurements made by antennas carried to high altitudes in aircraft, balloons and satellites show that the intensity of the radiation is isotropic to approximately one part in 1,000.

The homogeneity of the Universe is more difficult to express quantitatively. The Universe is of course highly inhomogeneous on a small scale such as that of the solar system or our galaxy. On a larger scale the homogeneity of its content of matter is indicated by the uniformity of the distribution of visible galaxies and radio sources and also by the uniformity of a universal background of X rays. On the scale of the entire observable Universe the most sensitive indicator of homogeneity is the universal background of microwaves, which is homogeneous on small angular scales to better than one part in 10,000. The measurement is not definitive, however, because the radiation might have been rescattered by the intergalactic medium on its journey to the solar system and hence its variations might have been smoothed.

The microwave background radiation establishes much more than the validity of the cosmological principle for the recent history of the Universe. The spectrum of the radiation is identical with the one that would be generated by a black body (a perfect emitter of radiation) at a temperature of 2.7 degrees Kelvin (2.7 degrees Celsius above absolute zero). The radiation is today only a feeble glimmer, but it attests to a fiery past. For the radiation to have the spectrum of a black body the early Universe must have passed through a hot, dense phase. One of the most remarkable predictions of modern cosmology was the suggestion made by George Gamow (and his co-workers Ralph A. Alpher and Robert Herman) that a remnant of the Big Bang should still be visible as a pervasive background of black-body radiation. Gamow believed all the chemical elements with the exception of hydrogen could have been created in the hot, dense phase of the Universe just after the Big Bang. To him the Universe was a giant fusion reactor, and the simple requirement that the Universe not immediately burn all its hydrogen into helium led directly to the prediction that the radiation, although greatly cooled and diluted by the expansion, should still be present with a temperature of about five degrees K.

Today it is known that the Universe did not stay hot and dense long enough for the heavy elements such as carbon and iron to be built up in successive reactions from primordial protons and neutrons. It is now known that the heavy elements were synthesized in the interior of stars. The oldest stars seem to be deficient in heavy elements, which means that at the time of their formation their environment was poor in such elements. Yet the helium abundance of the old stars is essentially the same as that of much younger stars rich in heavy elements. Moreover, many kinds of galaxies are alike in their helium abundance. The uniform universal distribution of helium, which is second only to hydrogen in its cosmic abundance, indicates that it was chiefly created not in stellar interiors but in the hot aftermath of the Big Bang, as Gamow visualized. The general agreement between his prediction of a black-body radiation of five degrees K and Penzias and Wilson's discovery of radiation of 2.7 degrees K and the accurate prediction of the primordial helium abundance are the most compelling arguments for a hot Big Bang.

From data on the microwave background radiation, theorists were able to calculate a new fundamental quantity: the ratio of the number of photons (the massless particles of electromagnetic radiation) in the Universe to the number of nucleons (the massy protons and neutrons). The ratio is about 10^8 photons to one nucleon and is a measure of the average thermal entropy associated with each nucleon. Entropy, usually represented by the letter S, is defined as a number that indicates how many states are possible in a system. To put it another way, entropy is a measure of randomness or disorder. A liquid is an example of a high-entropy system because its atoms can arrange themselves in a huge variety of ways; a crystal lattice is an example of a lower-entropy system because its atoms are arranged in a highly ordered way. The ratio of the density of photons to the density of nucleons averaged over a large volume of the Universe is a measure of the average entropy because photons constitute the most disordered states of thermal energy and nucleons constitute the most ordered states. Hence the relative abundances of these two kinds of extreme state are a measure of the average entropy.

According to the second law of thermodynamics, the total entropy of the Universe increases continuously as time goes on. This means that at the time of the Big Bang S was less than 10^8. The isotropic expansion of the Universe did not dissipate much heat, so that any increase in S must be due to other mechanisms. An entropy of 10^8 is quite large compared with the entropy of about 1 exhibited by systems in the terrestrial environment. This means that the Universe as a whole is a relatively hot place. A physicist's first reaction to a hot system with a high degree of regularity is to suppose much thermal energy was dissipated in the history of the system, the heat and the regularity being the aftermath of frictional smoothing processes. We shall try to demonstrate, however, that for the Universe this interpretation is unlikely to be correct.

By extrapolating backward from the present-day expanding Universe to the time before galaxies formed, cosmologists have traced the origin of the Universe to a singularity: a state of apparently infinite density. The singularity represents the origin of space and time perhaps 10 billion years ago. Before that time, the laws of physics known today did not apply. Does that testify to the physicist's knowledge or to his ignorance? Before 1965, cosmologists debated the physical significance of this singularity. Some theorists thought it might be a spurious manifestation of the particular coordinate system chosen to describe the dynamics of the expansion of the Universe, a manifestation similar to the singularity in geography. On a terrestrial globe, there is a coordinate singularity at the North and South poles, where the grid squares of longitude and latitude vanish as the meridians of the globe intersect. Yet the fabric of the world does not physically break down at the poles.

Since 1965, several theorists have independently shown that the cosmological singularity is not the result of a poorly chosen coordinate system. On the contrary, the singularity seems to be general, physically real and an inevitable consequence of the fact that gravity is attractive and acts indiscriminately on everything, including photons. The singularity was probably characterized by infinite density and curvature, although all that is known with certainty is that a material observer moving back into the past would experience an abrupt and disconcerting end to his trip through space-time when he encountered it. He would be unable to travel any farther because the laws of physics require the Universe to have a space-time boundary. (The traveler need not, however, experience the Big Bang with its infinite densities and temperatures.)

The cosmological singularity is similar to the singularity in the event horizon of a black hole, the hypothetical surface in which matter and light rays are confined by gravity. Nothing leaves a black hole because the velocity needed to escape the grasp of its gravity is greater than the speed of light, which the laws of physics require cannot be exceeded. It has been shown that an unfortunate astronaut who fell freely into a black hole would reach a physical singularity within a finite time (as time would be measured on his own watch). The singularity would be invisible to external observers since it lies inside the event horizon of the black hole. Now, since photons feel the pull of gravity, it is possible to determine whether at some time early in the history of the Universe the photons that currently make up the microwave background radiation could have exerted sufficient gravitational pull to create a trapped region analogous to a black hole. Within that region would be a cosmological singularity, which the physicist defines as the beginning of the Universe.

That the universal expansion originated at a singularity has a far-reaching consequence in terms of the points in the Universe that causally influence one another. Since no signal can move faster than the speed of light, an observer can be affected only by events from which a photon would have time to reach him since the beginning of the Universe. Such events are described as lying within the observer's horizon. Consider two points with a spatial separation x at the beginning of time. Before the time it takes for light to travel between them (x/c, where c is the speed of light), the points will not "see" each other, "know" of each other's existence or be able to affect each other in any way. In general, regions of the Universe with a spatial separation greater than ct will not know of each other's existence until a time t has elapsed. What this means is that regions of the isotropic microwave background in different directions of the sky (say more than 30 degrees apart) could never have casually influenced each other at any time in the past. That creates a paradox: How did causally disjointed primordial regions of the Universe come to have the same temperature and expansion rates today to within at least one part in 1,000?

A further twist to the conundrum of the origin of the uniform background radiation comes from the fact that although the Universe is quite regular on the scale of several tens of millions of light-years, it does have on a smaller scale some spectacular inhomogeneities in the form of galaxies and clusters of galaxies. The strength of the gravitational field exerted by the largest of these inhomogeneities suggests that their ancient precursors would have created an anisotropy in the microwave background radiation over a scale of a few angular degrees. Observational astronomers are currently searching for such an anisotropy. Anisotropy would also arise from the remnants of ancient directional disparities in the universal expansion rate. It is not clear what physical mechanisms have smoothed the irregularities into the structured Universe of today.

What cosmologists are trying to account for are the entropy and the large-scale structure of the Universe, which appear to have existed when the Universe was less than a minute old. Such epochs are best defined, however, not in temporal units such as minutes or years, which are subject to correction as the yardsticks of astronomy are refined, but in units of red shift, which express the amount by which the Universe was compressed with respect to its present size.

On the one hand, there are "chaotic" cosmologists, who, like biologists, maintain that the properties of the Universe are the result of evolutionary processes. They have tried to show that a kind of gravitational natural selection could deliver the present large-scale structure as the inevitable result of physical smoothing and heating processes that have been going on since the Big Bang. If it could be demonstrated that the present structure would have arisen no matter what the initial conditions were, then the uniqueness of the Universe would be established in theory as well as in actuality.

On the other hand, there are "quiescent" cosmologists, who appeal largely to the initial conditions to explain the present structure of the Universe. They hypothesize that when the Universe was created at the singularity, it had certain definite and preferred structural features for reasons, say, of self-consistency, stability or uniqueness. This means that gravitational evolutionary processes played a role not in shaping the overall configuration of the Universe but only in molding substructures such as galaxies, stars and planets. A good deal of the theoretical work in cosmology over the past decade has centered on finding ways of distinguishing between the two alternative cosmologies.

It is now believed that at ordinary temperatures all natural phenomena are governed by four fundamental forces: the gravitational force, the electromagnetic force, the weak force and the strong (or nuclear) force. These forces in conjunction with a small number of additional parameters such as particle mass determine the structural characteristics of the Universe. As the history of the Universe is traced backward in time from the present to the singularity at a red shift of infinity some 15 billion years ago, each of the four fundamental forces will at some point come to dominate the others. Today the gravitational force is the one that governs the dynamics of the large-scale expansion. Although the gravitational attraction of two protons is 10^{40} times weaker than their electromagnetic repulsion, gravity becomes increasingly important in a system with a huge number of particles such as the Universe. In the Universe the number of positively charged particles should be equal to the number of negatively charged particles, and so on the whole the attractive and repulsive electromagnetic interactions cancel out and exert no significant long-range forces over large regions. Since gravity is only attractive, however, it plays a major role in massive systems. It is the influence of the gravitational force, which has infinite range and acts on the photons of radiation as well as on the particles of matter, that determines the size of the largest objects in the Universe, such as planets, stars, galaxies and clusters of galaxies.

If a local region of the early Universe happened to have a density higher than that of the surrounding regions, it would gravitationally attract more matter than the less dense regions. As it contracted under the influence of its own gravity, it would increase in density and attract matter even more efficiently. What began as a fluctuation in a fairly homogeneous Universe would eventually snowball into a huge inhomogeneity. Galaxies seem to have formed at times equivalent to red shifts between 10 and 100 and to have come together in clusters at later times. It is not known with any certainty, however, whether galaxies formed out of fragments of much larger fluctuations that disintegrated or out of smaller fluctuations that came together because of their mutual gravitational attraction.

At a time equivalent to a red shift of 1,000 (about 300,000 years after the Big Bang), gas pressure is stronger than gravity over a dimension equivalent to that of about 100,000 suns, but gravity is much stronger than gas pressure over larger dimensions. Fluctuations on the order of these larger dimensions grow until they eventually become large enough to collapse and form bound objects spanning a range of masses from those of globular star clusters to those of galaxies. At times equivalent to red shifts greater than 1,000, the chief source of pressure is not gas pressure but the pressure of thermal radiation. At that epoch, the dynamical behavior of density perturbations is determined by the electromagnetic force.

Under these circumstances, photons form a viscous fluid that inhibits the movement of electrons and protons. An electron, for example, would scatter impinging photons, which it feels as electrical pulses. Because of the law of the conservation of momentum, the scattering would slightly alter the electron's trajectory. The net result is that the electron, locked into the radiation field by the ceaseless barrage of photons, cannot go anywhere. Once the electron joins a proton to form a hydrogen atom, however, it effectively no longer feels photons because it is interacting primarily with the electric field of the proton. The lack of movement of individual electrons means that fluctuations in the density of matter, called isothermal density fluctuations, are preserved until the time when electrons and protons combine to form electrically neutral atoms. Such atoms can travel freely through the radiation, and so the gravitational growth and collapse can proceed. The isothermal density fluctuations start to collapse when they acquire a mass of more than 100,000 suns. Many such fluctuations come together to form galaxies.

Objects could also form from another kind of perturbation, called an adiabatic fluctuation, in which matter and radiation are perturbed together. If matter and radiation were squeezed slightly in a confined space, the excess pressure would create a kind of sound wave. Yet just as sound waves in air eventually dissipate and fade, so would a primordial sound wave. The critical length and mass below which the wave would be damped completely is determined by the ability of photons to escape from the adiabatic fluctuation in the time since the beginning of the Universe. By a time equivalent to a red shift of 1,000, only adiabatic fluctuations more massive than the observed size of a massive galaxy or a group of galaxies survive the damping. In other words, less massive adiabatic fluctuations are smoothed out, whereas the more massive ones perhaps survive, grow and eventually collapse into massive galaxies and groups of galaxies.

In summary, this reconstruction of the probable evolution of the hot early Universe consists of two kinds of density fluctuation, the isothermal and the adiabatic, which roughly correspond respectively to a globular star cluster or a dwarf galaxy and to a cluster of galaxies. The picture is undoubtedly a gross oversimplification because in general an arbitrary inhomogeneity would be expected to have an admixture of both isothermal and adiabatic components. Moreover, newly formed structures could merge or fragment, leaving no trace of their previous individual identity. In spite of these qualifications, cosmologists would be rash to ignore this simple picture because the preferred masses that emerge out of the two kinds of fluctuation are comparable to the masses of the objects whose origin they are trying to explain.

At a time equivalent to a red shift of 10^{10}, when the Universe was only a few seconds old and its temperature was 10^{10} degrees K, physical processes are mediated by the weak force. This force governs certain radioactive decay processes involving a free neutron or a neutrino: a spinning pointlike particle with no charge and negligible mass. At times equivalent to red shifts greater than 10^{10} the weak force keeps the protons and neutrons in thermal equilibrium: a statistical state where the number of particles with properties (position, mass, energy, velocity, spin and so on) in a specified range remains constant because the rate of particles entering that range is equal to the rate of particles leaving it. When particles achieve a state of thermal equilibrium, their behavior is determined not at all by their history but entirely by a set of statistical laws based on their temperature. This means that it is unnecessary to pry any farther into the past to understand the behavior and the relative concentrations of protons and neutrons.

The weak force in conjunction with the Big-Bang model of expansion leads to the prediction that the abundance of primordial helium in the Universe is between 25 and 30 percent. This prediction, which has now been confirmed, has led to a precise and direct observational test of the Big-Bang cosmology. What the success of the prediction means is that a few seconds after the Big Bang, the Universe was at least as regular on a large scale as it is today and had almost the same entropy. Cosmologists are trying to determine the degree of regularity before that time.

At times equivalent to red shifts greater than 10^{10}, neutrinos and their antiparticles (antineutrinos) play a big role. Today these particles are quite elusive because they almost never interact with anything in the rarefied medium of the present Universe. Yet when the Universe was a little less than a second old, matter and radiation were so dense that neutrinos rapidly interacted strongly with them. At a time equivalent to a red shift of 10^{10}, a typical neutrino traversed a significant extent of the Universe before it collided with another particle. This means that the neutrinos could effectively transport energy and momentum over extremely large distances. They would do so by absorbing energy in high-energy regions of the Universe and transferring it by occasional collisions to low-energy regions. As a result, the neutrinos act to smooth out any nonuniformities in the distribution of matter that might have been created by directional differences in the overall expansion of the early Universe.

The possibility of smoothing by neutrino-transport processes was suggested in 1967 by Charles W. Misner of the University of Maryland in the hope that such processes could remove a host of irregularities associated with the initial singularity of a chaotic Universe. This hope has now foundered on the realization that if the initial expansion were sufficiently asymmetrical, the Universe would expand so rapidly that there would not be enough time for neutrinos to collide with other particles. In other words, a highly anisotropic universe could remain that way. It seems that neutrino transport and other smoothing processes could only remove anisotropies below a certain level. No matter how efficient a smoothing process was postulated, it was always possible to imagine a model universe that would nonetheless remain much more irregular than the present-day Universe actually is. This consequence of the weak force cannot satisfy the central tenet of chaotic cosmology: the evolution of the regular present-day Universe from any initial state no matter how irregular.

Perhaps the homogeneous Universe could be the result of more complex smoothing processes, such as Misner's ingenious "mixmaster" Universe in which matter would be mixed periodically by explosions expanding at the speed of light first in one direction and then in another. This process is an exotic and complex one developed on the basis of the general theory of relativity. The probability of such explosions, however, seems to be infinitesimal. In fact, the conditions required for such mixing are almost as special as those required for the initial Universe to be precisely regular. This means that the mixmaster model explains very little. The search for complex smoothing processes that would require fewer special conditions is part of the continuing work in chaotic cosmology.

It is time to leave the epoch of the weak force to move closer to the cosmological singularity and tentatively probe the first milliseconds of the Universe. There, where the temperatures and the particle energies exceed those achieved by the most powerful man-made accelerator and the radiation density is comparable to the density of the atomic nucleus, the dominant force between particles is the strong force. The extrapolation of our model into these earliest times is somewhat precarious because the understanding of the basic physics is not complete. When the strong force annihilates a proton and an antiproton, it gives rise to two energetic photons moving in opposite directions. In the first millisecond of the Universe, the temperature would have been so high that such an annihilation and the inverse process, the spontaneous production of nucleons and antinucleons from photons, would be quite efficient. Nucleons and radiation would have been indistinguishable. The average entropy of 10^8 photons per nucleon observed today implies that when the last annihilation took place, one proton survived for every 10^8 photons created by the destruction of other pairs of particles and antiparticles.

It seems that just before the Universe was a millisecond old there was a minute imbalance between matter and antimatter: 1.00000001 particles per antiparticle. Until recently, the origin of this peculiar imbalance was a complete mystery because of a principle having to do with baryons: heavy particles, including nucleons, that feel the strong force. Physicists believed the number of baryons in a system minus the number of antibaryons was absolutely fixed for all time. No interactions or transformations of the particles could ever change this quantity. If this were true for the Universe as a whole, the asymmetry of one part in 10^8 of matter over antimatter must have been built into the initial structure of the Big Bang.

Over the past decade cosmologists have been actively investigating the consequences of a new extension of the theory of matter in which the electromagnetic, the weak and the strong force are all unified at sufficiently high temperatures. Above 10^{26} degrees K these forces lose their individuality, whereas at lower temperatures they seem to be independent (although they are actually different aspects of an underlying unity). This kind of unification is possible only if the quarks that make up protons and other elementary particles are able to decay. Such decay has a surprising consequence: the proton is an unstable particle, although it has an average lifetime of about 10^{31} years. In spite of the infrequency of proton decays, physicists hope to observe some decay within the next few years by an experiment examining a sufficiently large mass: 1,000 tons of matter consisting of roughly 10^{32} protons.

The possibility of such unusual processes indicates that the level of particle-antiparticle asymmetry in the Universe, which determines the observed entropy, is not absolutely invariant. It can change dramatically in the first 10^{-35} second of the Universe, when the processes that mediate the decay of protons are abundant. Recent work shows that after this early instant a stable level of asymmetry between particles and antiparticles is eventually frozen into the Universe, and the predicted value is close to the observed asymmetry of one part in 10^8. The observed entropy need not be strongly dependent on the initial conditions of the Big Bang in the first 10^{-35} second.

So far we have been taking into account only the corpuscular properties of matter. As we speculate on what happened right after the Big Bang, the wave properties of matter must also be considered. According to quantum mechanics, every particle behaves as a wave with a length equal to 2.1×10^{-37} divided by the particle's mass. This wavelength, called the Compton wavelength, is infinitesimal by everyday standards, but only 10^{-23} second (the Compton time) after the Big Bang, the Compton wavelength (10^{-13} centimeter) of a proton would be equal to the size of the causally connected region of the Universe. There is nothing fundamental about this scale, however, because there is nothing fundamental about protons, which are of course made up of quarks.

The ultimate barrier is reached at times close to 10^{-43} second (the Planck time) after the Big Bang, because causally connected regions of the Universe were compressed to a scale smaller than the Compton wavelength of their entire mass content. Before the Planck time, the usual interpretation of space-time is probably invalid because quantum-mechanical fluctuations dominate the geometry of space-time. Undoubtedly many secrets of the Universe would be revealed by an understanding of pre-Planck time, but achieving such an understanding is currently a remote possibility. Achieving it will probably have to await a new physical theory that synthesizes the theory of relativity and quantum theory. Cosmologists now regard the Planck time as being in effect the moment of creation of the Universe; they leave to speculation any possibility of an earlier phase of evolution.

The epoch between the Planck time and the Compton time is a little more accessible to theoretical work. In that epoch, quantum mechanics points to a mechanism that might have erased irregularities in the Universe. According to quantum mechanics, all space is filled with pairs of "virtual" particles and antiparticles. Such particles, which materialize in pairs, separate, come back together and annihilate each other, are called virtual because unlike real particles they cannot be observed directly by particle detectors, although their indirect effects can be measured. If a pair of virtual particles is subject to a force field that is either extremely powerful or rapidly varying, its components might separate so quickly that they could never come back together. In this case the virtual particles would become real ones, their mass being supplied by the energy of the force field. Close to the Planck time, the required force field might be generated by the changing expansion dynamics of the Universe itself.

The Russian astrophysicist Ya. B. Zel'dovich has proposed that the production of real particles from virtual ones would erase the anisotropies and nonuniformities in the initial structure of the Universe. It is imagined that radiation and particles would be preferentially spawned in the overly energetic regions of space. The newly formed particles would transfer energy from high-density regions to lower-density ones. Moreover, they would tend to equalize the rates of expansion in different directions. Much as people stepping onto a rotating merry-go-round tend to slow it down, the sudden appearance of particles would slow down the rapidly spinning and moving regions. Perhaps it is this quantum-physical mechanism that is responsible for the homogeneous large-scale structure of the present-day Universe.

In the above discussion, we have reviewed several smoothing mechanisms that might be responsible for the present regularity of the Universe, mechanisms such as neutrino transport and the creation of real particles from virtual ones. As promising as these mechanisms may seem, we believe thermodynamical considerations demonstrate that they are severely limited. Since entropy increases with time, the present level of entropy in the Universe (10^8) puts an upper limit on the amount of dissipation that has occurred during the past history of the Universe. The smoothing of anisotropies and inhomogeneities in the initial structure would irreversibly convert the energy of irregularity into heat energy. The erasure of primordial chaos would generate radiation, but the low level of the microwave background radiation means that there could not have been an arbitrarily large amount of heating and smoothing in the past.

Moreover, it has been shown that in general the sooner after the Big Bang the irregularities are damped out, the more thermal radiation would be generated. Most of the dissipative mechanisms we have discussed would be fully operative right after the Big Bang. This means that if the Universe were made with anything but a small degree of irregularity, entropy would be created at the singularity in excess of the observed level. But how small must the initial irregularity have been? The work of Barry Collins and Stephen Hawking of the University of Cambridge shows that a highly but not perfectly regular Universe is unstable. The slightest deviation from regularity would tend to grow in time as the Universe expanded regardless of the dissipative mechanisms. To put it another way, a Universe beginning its expansion in anything but a precisely regular configuration would tend to become increasingly irregular. This means that the initial irregularities could only have been infinitesimal. Our own work indicates that there is a hierarchy of irregularities in the Universe. As progressively larger volumes of space are examined, the degree of irregularity decreases in a way that suggests the initial irregularities were only statistical fluctuations of a regular state.

What we are saying is that the present entropy level shows that the Universe has evolved in an exceedingly regular way from as far back as the first 10^{-35} second of its existence. Before that time, complex processes involving the nonconservation of the symmetry of particles and antiparticles and the quantum properties of the gravitational field erased any memory of the initial entropy per nucleon. Although the dissipation of chaos at those early times could also have generated many photons, nucleons would have been created as well, and the net number of photons per nucleon might have decreased or increased. Of course, the total entropy of the matter and radiation in the Universe must always increase.

An unsettled question is whether the initial conditions are unique or whether another set of conditions could have done the same job. Several candidates for initial conditions have been proposed by theorists. One possibility is that the strong force served to keep matter highly rigid at the time of the singularity. At extremely high densities two nucleons could repel each other in the same way that two like magnetic poles do. Such a repulsion, which would prevent heavy nuclei from collapsing, might come to dominate the overall behavior of the interacting particles. An early state dominated by the strong force would remain quite regular because the high pressure would prevent distortions or turbulence from developing as the density increased.

Another possibility, developed by Roger Penrose of the University of Oxford, is based on the proposal that the overall gravitational field of the Universe itself has an entropy that is proportional to and dependent on its uniformity. Since gravitational entropy, like all other forms of entropy, should always increase with time, the initial state of the Universe would have been one of low gravitational entropy and regularity. To minimize gravitational entropy, the Universe might have acquired a regular, isotropic configuration, as a soap bubble minimizes the entropy associated with its surface area by assuming an isotropic spherical shape. As the Universe ages and expands, the gravitational entropy would increase to reflect the growth of such inhomogeneities as galaxies and clusters.

Still another speculative possibility for the initial conditions of the Universe is based on Mach's principle: The motion of an object is determined not by the characteristics of some "absolute" geometrical space but solely by the material content of the Universe. (The principle was first advanced in 1893 by Ernst Mach in a critique of Newton's concept of an absolute space to which the motion of all objects is referred.) Although Mach's principle is not a consequence of the theory of relativity, it has been introduced into the theory either as a boundary condition on the relativistic equations or as a sieve for eliminating solutions that are not physically acceptable. Derek Raine of the University of Leicester has developed a detailed version of the latter alternative, and out of his work comes the requirement that the infant Universe be almost completely isotropic and homogeneous.

The last possibility we shall discuss is based on the idea that the initial conditions are limited by the very fact that they have led to human life on the earth. This idea, called the anthropic cosmological principle, was introduced by G. J. Whitrow of the University of London, Robert H. Dicke

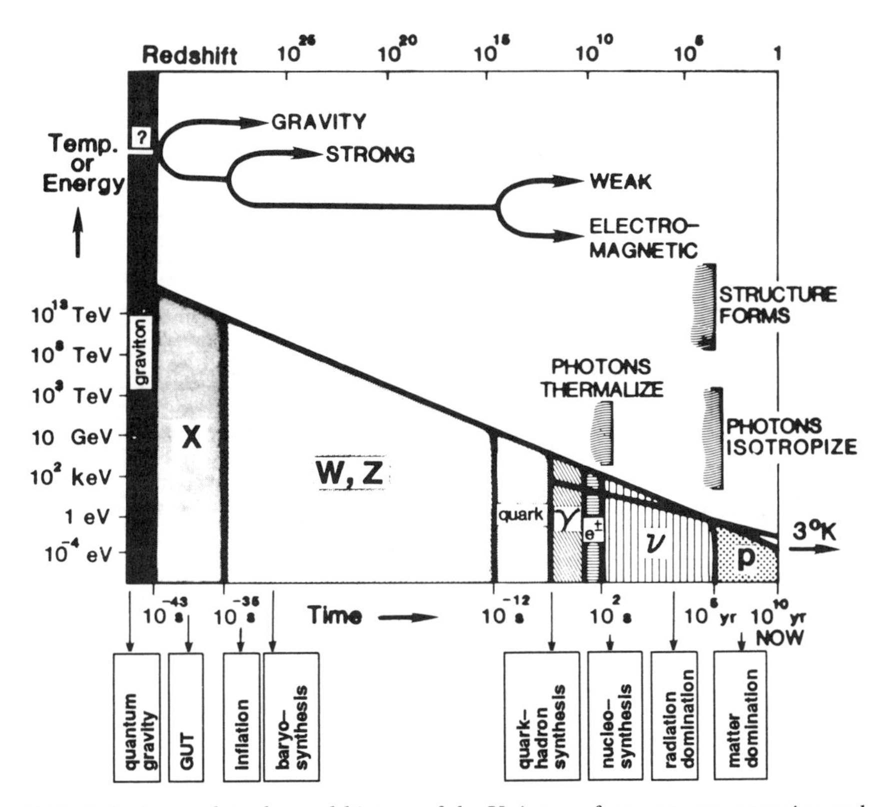

FIGURE 1. *A complete thermal history of the Universe, from extreme energies and exotic particles in the primordial fireball, to ordinary matter and the present radiation temperature of 3 degrees Kelvin. The temperature of the Big Bang decreases with the continuing expansion of space.*

of Princeton University and Brandon Carter of the Meudon Observatory and was developed by John A. Wheeler of the University of Texas at Austin. Consider how this principle bears on the question of the size of the Universe. Since the Universe is constantly expanding, its size depends on its age. The anthropic cosmological principle has convinced us that the Universe must inevitably be about 10 billion light-years in diameter. A smaller Universe would have existed for less than the billion years necessary for the heavy elements essential to human life to be synthesized by thermonuclear reactions in the interior of stars. Moreover, if the Universe were much bigger and hence much older, the stars needed to establish the conditions of life would have long since completed their evolution and burned out.

COSMIC TIME	EPOCH	RED SHIFT	EVENT	YEARS AGO
0	SINGULARITY	INFINITE	BIG BANG	15×10^9
10^{-43} SECOND	PLANCK TIME	10^{32}	PARTICLE CREATION	15×10^9
10^{-6} SECOND	HADRONIC ERA	10^{13}	ANNIHILATION OF PROTON-ANTIPROTON PAIRS	15×10^9
1 SECOND	LEPTONIC ERA	10^{10}	ANNIHILATION OF ELECTRON-POSITRON PAIRS	15×10^9
1 MINUTE	RADIATION ERA	10^9	NUCLEOSYNTHESIS OF HELIUM AND DEUTERIUM	15×10^9
1 WEEK		10^7	RADIATION THERMALIZES PRIOR TO THIS ERA	15×10^9
10,000 YEARS	MATTER ERA	10^4	UNIVERSE BECOMES MATTER-DOMINATED	15×10^9
300,000 YEARS	DECOUPLING ERA	10^3	UNIVERSE BECOMES TRANSPARENT	14.9997×10^9
1×10^9 YEARS		10	GALAXIES BEGIN TO FORM	14×10^9
7×10^9 YEARS		1	GALAXIES BEGIN TO CLUSTER	8×10^9
1×10^9 YEARS		10	OUR PROTOGALAXY COLLAPSES	14×10^9
1×10^9 YEARS		10	FIRST STARS FORM	5×10^9
2×10^9 YEARS		5	QUASARS ARE BORN; POPULATION II STARS FORM	13×10^9
7×10^9 YEARS		1	POPULATION I STARS FORM	8×10^9
10.2×10^9 YEARS			OUR PARENT INTERSTELLAR CLOUD FORMS	4.8×10^9
10.3×10^9 YEARS			COLLAPSE OF PROTOSOLAR NEBULA	4.7×10^9
10.4×10^9 YEARS			PLANETS FORM; ROCK SOLIDIFIES	4.6×10^9
10.7×10^9 YEARS			INTENSE CRATERING OF PLANETS	4.3×10^9
11.1×10^9 YEARS	ARCHEOZOIC ERA		OLDEST TERRESTRIAL ROCKS FORMS	3.9×10^9
12×10^9 YEARS			MICROSCOPIC LIFE FORMS	3×10^9
13×10^9 YEARS	PROTEROZOIC ERA		OXYGEN-RICH ATMOSPHERE DEVELOPS	2×10^9
14×10^9 YEARS			MACROSCOPIC LIFE FORMS	1×10^9
14.4×10^9 YEARS	PALEOZOIC ERA		EARLIEST FOSSIL RECORD	600×10^6
14.55×10^9 YEARS			FIRST FISHES	450×10^6
14.6×10^9 YEARS			EARLY LAND PLANTS	400×10^6
14.7×10^9 YEARS			FERNS, CONIFERS	300×10^6
14.8×10^9 YEARS	MEZOZOIC ERA		FIRST MAMMALS	200×10^6
14.85×10^9 YEARS			FIRST BIRDS	150×10^6
14.94×10^9 YEARS	CENOZOIC ERA		FIRST PRIMATES	60×10^6
14.95×10^9 YEARS			MAMMALS INCREASE	50×10^6
15×10^9 YEARS			HOMO SAPIENS	1×10^5

Major events in the Universe's history are listed. The time scale of ancient events must be regarded as tentative because the precise age of the Universe is not known. It is therefore better to date such events in terms of the red shift, which is a measure of the degree of compression of the expanding Universe. At the ultra-high velocities characteristic of times close to the Big Bang, the red shift is equal to $(1 + v/c)/(1 - v^2/c^2)^{1/2} - 1$*, where* v *is velocity of the radiation source and* c *is velocity of light* (3×10^8 *meters per second).*

The anthropic cosmological principle bears on the Universe's entropy level of 10^8. If this number were increased by a factor of 1,000 or so, it would not be possible for protogalaxies to condense to the density at which stars would form. Without stars, the solar system and the heavy elements of living matter would not have been created. If the Universe were initially fairly irregular, it would have irreversibly generated copious quantities of heat radiation because of the many efficient entropy-generating channels open to it at the time of the singularity. Again this would have resulted in an entropy and a radiation pressure in excess of the values favoring the condensation of protogalaxies. Such a Universe could not have been observed by us. These ideas indicate one thing: man's existence is a constraint on the kinds of universes he could observe. Many features of the Universe that are remarkable to ponder are inevitable prerequisites of the existence of observers.

This unusual approach has even been extended to the values of the fundamental constants of nature. Consider a change in the strength of the strong-interaction coupling constant of only a few percent. As small an increase as 2 percent would block the formation of protons out of quarks and hence the formation of hydrogen atoms. A comparable decrease would make certain nuclei essential to life unstable. By the same token small changes in the electric charge of the electron would block any kind of chemistry and rule out the existence of stable planet-supporting stars.

Although the anthropic cosmological principle indicates why the structural features of the Universe are in some sense inevitable, it leaves the reason for these features a mystery. Whatever the scientific status of the anthropic cosmological principle may be, its impact on the history of ideas may be significant. The principle overcomes the traditional barrier between the observer and the observed. It makes the observer an indispensable part of the macrophysical world.

In 1980, Alan Guth proposed the theory of inflationary cosmology. A brief period of exponential expansion some 10^{-35} second after the Big Bang resulted in a universe that was incomparably smoother, incomparably more isotropic, than anything that previously existed. A wrinkled balloon, greatly inflated, had shed its wrinkles. Whether inflation is generic to the Big Bang is unknown. Detection of infinitesimal, large angular scale fluctuations in the cosmic microwave background supports, but does not verify, inflation. If inflation occurred, however, many types of possible initial conditions were surely shrouded from our view. Some of the goals of the chaotic cosmology program may finally have been vindicated, although it is by no means clear that arbitrarily chaotic and anisotropic initial conditions would have been inflated away.

GALAXY FORMATION

Formation of the Galaxies

Origin

Galaxies are the fabled island universes that Kant hypothesized a century ago. Stars numbered by the billions coexist in their mutual gravitational field, aimlessly orbiting the galaxy. It will take the sun about two hundred million years to complete even a single orbit around the Milky Way. The solar system has already been 20 times around the galaxy, and may have formed on the far side of our galactic system, perhaps within a spiral arm. Spiral arms are density waves, regions of compression that alternate with rarefactions and propagate in a spiral pattern through the orbiting stars and gas clouds that constitute the differentially rotating galaxy. Within the spiral arms, interstellar gas clouds accumulate in a frenzy of gravitational interaction, to collapse and form stars. These are regions of intense star formation and, inevitably, star death. Perhaps the chance compression of an interstellar cloud by impact with the remnant of an exploding star triggered the event that formed the sun some four and a half billion years ago.

The interstellar gas cloud, no different from those we see today throughout the Milky Way, was pushed over the precipice of gravitational instability as the compression induced it to collapse. The density mounted and the gas cooled more and more rapidly. The ability of the gradients in gas pressure to withstand the overwhelming force of gravity inevitably declined as the gas cooled. This only served to accelerate the collapse until density fluctuations in the cloud were amplified to the point that the cloud began to break up into many fragments. Many cloudlets formed, each destined to contract into a protosun. Thus was our sun born, into a cluster of stars that has long since dispersed around the galaxy. Such events were repeated a millionfold in the remote past during the early evolution of our galaxy. Indeed, the formation of stars continues to the present day, albeit at a reduced rate as the raw material of interstellar gas is slowly depleted.

Long ago, there was no galactic disk and there were no stars. The brilliant Milky Way, with its myriad points of light, did not exist as we see it today. Its ancestor was a huge cloud of almost pristine hydrogen, some 100 billion times the mass of the sun. The cloud, of which 25 percent by mass was helium, was itself a relic from the beginning of the Universe. The helium content is the vital clue: produced in the first three minutes of the Big Bang, it is a universal constituent of all ordinary matter whether in nearby stars or remote galaxies.

As matter expanded from the dense, hot phase of the beginning of the Universe, it condensed into the massive clouds that are the precursors of galaxies. Each protogalaxy contracted under its own gravity as it gradually radiated away the internal energy of its atoms. The cooling soon became a runaway process, as slightly overdense regions cooled more efficiently than their surroundings. The ensuing fragmentation led to a swirl of orbiting gas clouds. As the clouds collided with each other, their orbital energy turned into heat and was lost as radiation, the clouds gradually settling into a rotating gas disk only after billions of years had elapsed. The protogalaxy had acquired an angular momentum as a consequence of gravitational torques exerted from similar nearby protogalaxies. The angular momentum, being retained during the contraction, was responsible for formation of the disk of the galaxy. The clouds in turn contracted and fragmented to form the first clusters of stars. Over time, many of the star clusters dissolved because of the disruptive gravitational forces exerted by other clouds, and a galaxy emerged that resembled the Milky Way. A rotating disk of stars and gas clouds was the predominant feature, along with a compact central spheroid of stars that developed from those infalling gas clouds with very low angular momentum. Five billion more years elapsed before one of these interstellar clouds provided the womb from which our solar system eventually emerged.

The Cosmological Setting

The Universe is now about fifteen billion years old. For the first million years, it was dense, hot and opaque to radiation. But a curtain lifted as the hydrogen plasma cooled and condensed into hydrogen atoms. The cosmic microwave background is our direct view of the Universe at this primeval stage. This blackbody radiation is the descendant of the cosmic fireball that once filled the Universe, gradually cooling as the Universe expanded and is now observed as an isotropic blackbody radiation field in the microwave region of the spectrum. Once, however, the radiation field consisted of energetic photons, capable of ionizing hydrogen and maintaining matter in the form of a primeval plasma of protons and electrons.

During the plasma phase, the ions were strongly coupled to the radiation, and any fluctuations in density remained small. Once the radiation had cooled sufficiently, the protons and electrons combined to form hydrogen atoms about a million years after the Big Bang. Only then did the Universe become transparent to the radiation, and the matter subsequently suffered no resistance to the motion induced by local variations in the gravity field, which were due to the presence of the density fluctuations. Gradually, matter began to agglomerate around the largest primeval density fluctuation peaks. Large peaks grew larger, at the expense of lesser density peaks, as the force of gravity gradually asserted its influence on the expanding matter of the Universe. Eventually, enough matter collected for the gravity force to locally overwhelm the expansion, and clouds of gas formed. The minimum mass such clouds could have had was about one million suns, since only with a mass this large or greater could the pressure of the gas be restrained by self-gravity of the cloud. The smallest dwarf galaxies have masses in this range, and they may indeed be relics of the first generation of gas clouds. Clouds merged and grew in mass, and galaxy-mass clouds, or protogalaxies, must have formed in great numbers by the time the Universe had aged about a billion years.

The First Stars

The first stars to form were massive, short-lived stars. Within a few million years, such stars explode to become supernovae. Brighter at first than a billion suns, supernovae fade rapidly, to leave behind a rapidly spinning neutron star surrounded by a cocoon of gas that expands at nearly 10 percent of the speed of light. The gas contains the outer envelope of the exploding star and is highly enriched in many of the heavy elements that were synthesized in the core of the star before it exploded. The expanding shell plows into the surrounding interstellar gas, decelerates as it sweeps up more and more gas, and eventually breaks up, releasing the polluted matter into the interstellar medium. Each succeeding generation of stars to condense out of interstellar gas clouds is progressively enriched. Five billion years elapsed before a level of nuclear enrichment was attained that is found in the sun. Only then did the sun form, out of a cloud which consisted predominantly of hydrogen but with a two per cent admixture of elements heavier than helium.

One traces the chemical history of the galaxy by studying the abundances of various elements in stars of differing ages. In this way, astronomy comes to resemble archaeology as one searches for the oldest, rarest stars, inevitably found near the sun because this is the only region that can be probed in sufficient depth. Very few stars of extremely low metal content are found. This is how we infer that most of the first generation of stars were short-lived, and therefore massive.

This conclusion would not be very secure if our only datum was the infrequency of occurrence of such stars, since one can readily envisage other reasons for their rarity: for example, most of the matter in the galaxy might have been accreted after the first generation of stars had evolved. However, studies of the content of different elements, most notably oxygen and iron, have helped elucidate the past history of star formation. Oxygen is known to be a product of massive stars that weigh more than 20 solar masses; it is spewed out into the interstellar medium when these stars explode as supernovae. Iron is also synthesized in supernovae, particularly in the supernovae that arise from low mass stars, or white dwarfs, that undergo mergers. Observationally, however, one finds that oxygen, as well as other products of massive stars such as silicon and magnesium, is closely coupled in abundance to iron. Very little dispersion is found over the history of the disk, at least in the solar vicinity. One interpretation is that the mass distribution of stars in the early galaxy was weighted towards massive stars, which were then the dominant producers of iron as well as of oxygen. This dominance faded with time, and evidence in favor of this relatively recent change is inferred from the gradual decrease in the ratio of oxygen to iron with time seen in disk stars. It is also possible that this is a natural aging effect: only with time do the low mass stellar producers of iron make their full contribution to the galactic iron budget.

Searching for Protogalaxies

The holy grail of cosmology is the elusive protogalaxy, a galaxy in the process of formation. The period during which the parent cloud contracted and formed the first generation of stars was about a billion years, a time-scale that may be inferred from the observed age spread of the oldest star systems in the galaxy, globular star clusters. During this period, the rate of star formation must have exceeded the present rate by a large factor, of 10 or more. The protogalaxy was highly luminous. As we look back into the most remote regions of the Universe, a natural expectation is that one should see space teeming with protogalaxies. Hitherto, this expectation has not been realized.

Astronomers have found some exceedingly curious and rare candidates that may well be protogalaxies. These distant galaxies are extremely luminous and are observed to be glowing with radiation from highly ionized, predominantly hydrogen plasma. They are gigantic counterparts of ionized hydrogen clouds in our own galaxies, known to be excited by massive, newly formed stars. Perhaps buried in these remote gas clouds are the first stars in a forming galaxy.

But equally likely is that a buried quasar, the nucleus of a future galaxy, is the source of the emitted light. Quasars are the most luminous objects in the Universe, and are believed to be a transitory stage associated with the birth of a galaxy. The innards of a collapsing gas cloud can attain high density and form a giant black hole, which emits prolific amounts of energy as it accretes surrounding material. When the supply of gas and nearby stars is exhausted, the central region becomes the innocuous nucleus of the surrounding galaxy.

An alternative strategy to search for forming galaxies is to take the deepest possible images of the darkest regions of the sky. Here, one is peering out of the galaxy, and as one takes longer and longer exposures, one is probing even further back in time. The faintest images detectable correspond to the light from a candle on the moon. The results of such searches have led to the discovery of a new, numerous population of dim blue galaxies that cover much of the sky. The galaxies are so faint that we cannot ascertain distance: they remain an enigma, and may be our best prospect for the elusive protogalaxies.

Galaxies In Collision

Much can be learned about how galaxies formed by studying nearby galaxies. Processes that are rare today may have been common earlier in time when galaxies were forming, and these rare phenomena can give invaluable clues about cosmic evolution. Mergers between galaxies appear to trigger a violent outburst of star formation. A close encounter between two galaxies will, if at sufficiently low speed, inevitably lead to a merger.

Gas clouds within the galaxies undergo collisions, even though stars do not. Gas cloud collisions result in loss of orbital energy of the gas clouds. The gas interactions are boosted by the tidal forces exerted as the stellar components merge together. The result is a massive, dense cloud of gas in the centre of the merging galaxy system. Such a cloud is unstable and must collapse and fragment into many stars.

The resulting episode of star formation is observed in the most luminous star-forming galaxies, which inevitably are in the final throes of a merger. Such objects are rare today, because the phenomenon is short-lived, but relatively common is the occurrence of the after-effects of such mergers. Many apparently normal nearby elliptical galaxies are found on close inspection to reveal signs of a merger in their distant past. The evidence is in the form of very faint shells of stars seen at the outermost periphery of many ellipticals. Analogous to the ripples in a pond that persist after a stone is thrown into the water, the disturbances in the gravity field slowly fade. Over billions of years, however, stars accumulate in the ripples in the gravitational potential, and mark the transient occurrence of a merger that, in the central regions, has produced what to the casual observer looks like a normal elliptical galaxy. Other features, such as dust lanes or knots of younger stars, are occasionally found that also bear witness to a more violent past.

Computer simulations have succeeded in demonstrating that a merger between two gas-rich spiral galaxies, the most common galactic systems in the Universe, results in the formation of an elliptical galaxy. The elliptical is characterized by a smooth, centrally concentrated, spheroidal distribution of old red stars, whereas its precursors are disk galaxies with patchy dust lanes and many clusters of young, blue stars. The merger greatly accelerates the rate of cloud formation, aggregation and collapse, exhausting almost all of the gas reservoir in a violent outburst of star formation, similar to that observed in the most luminous infrared galaxies that are identified as starburst galaxies. In many instances, one can identify the actual merging systems.

The Cosmological Hierarchy

While mergers are rare today, perhaps one galaxy in a hundred undergoing such strong tidal interactions, cosmology strongly suggests that mergers were a common phenomenon in the past. Structure developed in a hierarchy, from small to large scales. The most massive objects in the

Universe, clusters and superclusters of galaxies, are just forming at the present epoch. Studies, especially maps at X-ray wavelengths of the hot intracluster gas, have revealed that many clusters of galaxies are morphologically young systems. The gas has a clumpy distribution: eventually, when complete dynamical mixing has occurred, it will appear uniform, spherically symmetric and centrally concentrated, as indeed is observed for mature galaxy clusters.

The fact that on megaparsec scales today we are witnessing gravitationally driven aggregation and merging of structure provides an important clue. On smaller scales, corresponding to those of individual galaxies, structure formation must have occurred by a similar process in the remote past. When the Universe was perhaps a billion years old, a tenth or less of its present age, merging of dwarf galaxy mass clouds was a frequent phenomenon that led to the emergence of individual massive galaxies, surrounded by relic dwarf satellites. Our own Local Group is such an aggregation, dominated by the Milky Way and the Andromeda Galaxy, accompanied by a handful of dim dwarfs.

This suggests that dwarfs are the building blocks of giant galaxies. Indeed, the abundant population of faint blue galaxies are almost certainly such dwarfs undergoing a transient burst of star formation. One expects the nearby universe to be littered with the fossilized relics of such early star formation episodes. Exceedingly dim nearby galaxies are indeed found that are barely discernible above the night sky. These are most likely the failed counterparts of today's bright galaxies. Nature most likely fails many times before it succeeds, and dim dwarfs and giants may be testimony to this process. Indeed, giant gas clouds are found in the distant universe, detectable because they absorb light in the spectral lines of a few abundant elements from an even more distant quasar. Such absorbing gas clouds cover about 20 percent of the sky, as a transparent cosmic wallpaper. These clouds most likely contracted and formed stars, to evolve into Milky Way-type galaxies, occasionally into giants—dim, low-surface-brightness galaxies where gravity has never been sufficiently strong to provoke vigorous star formation. Only by studying all of the denizens of the cosmic zoo can we ever hope to understand the complex processes that led to galaxy formation.

The Cosmological Conundrum

Density fluctuations were created at the threshold of cosmic time. The cosmic microwave background with its near perfect blackbody spectrum at the observed temperature of 2.73 degrees Kelvin bears witness to the early hot phase of the Big Bang. This encourages cosmologists to extrapolate the theory of the Big Bang back in time to within 10^{-43} sec of the initial singularity, the instant at which classical physics, as opposed to the as yet poorly understood theory of quantum gravity, first becomes applicable. Quantum fluctuations were inevitably present at this epoch, when the fundamental scale of elementary particles was of the same order as the gravitational radius that demarcates the strong curvature of space. These fluctuations were boosted in scale, from microscopic to macroscopic dimensions, by an inflationary phase of the expansion of the Universe that occurred some 10^{-35} sec after the Big Bang. From these fluctuations, all structure subsequently evolved. The primeval fog lifted, a million years after the Big Bang, and density fluctuations began to grow, to eventually evolve into protogalaxies. Before then, the radiation field of the Universe inexorably held the primeval plasma in its unrelenting grip until neutralization of the primordial plasma allowed hydrogen atoms to accumulate into the great clouds that were the precursors of galaxies.

The ultimate test of such a model is the prediction of temperature fluctuations that are close to, but just intriguingly below, the latest limits on the smoothness of the cosmic microwave background. Its uniformity in the sky testifies to the near homogeneity of the very early Universe. What is remarkable is the extreme degree of uniformity observed in the cosmic microwave background radiation. Very recently (1992), blotches, hot spots or cold spots, have been detected by the COBE satellite over 10 degree scales at a level of 30 microkelvin, or about 0.001 percent. The Universe, a million years after the Big Bang, is inferred to have been almost completely, but not quite, homogeneous. The homogeneity cannot have been perfect; otherwise no galaxies would have formed. Infinitesimal density fluctuations were present in the primordial Universe with an amplitude of about 10^{-5}, precisely with the amplitude required (to within a factor of 2!) in order for galaxies to have formed by the present epoch via growth by gravitational instability. This is the current paradigm to which cosmologists subscribe.

Confirmation of the detection of these primordial fluctuations in the cosmic fireball on smaller angular scales is urgently awaited: only on degree scales would one hope to directly detect the precursor seeds from which the observed large-scale structure in the galaxy distribution arose. This is a golden age in cosmology: we are on the threshold of confirming whether our present ideas about cosmology and the origin of large-scale structure have any validity.

Slow-motion Galactic Birth

The timescale over which primordial material collapsed to form our Galaxy is of considerable cosmological significance. The debate is hinged on whether the collapse occurred over a 'dynamical' timescale, which would require all the old stars in the Galaxy to have formed within an initial free-fall time of less than a billion (10^9) years, or whether the early collapse was a more gradual process. Perhaps our Galaxy contracted over several billion years: if this is the case, there are noteworthy, even revolutionary, implications for our picture of galaxy formation. Observations of two globular star clusters by R. J. Dickens et al., reported in May 1991, strengthen the view that there is a spread of about 3 billion years in the ages of some of the oldest stars in our Galaxy.

The significance of globular clusters, gravitationally bound associations of up to a few million stars, is that they are thought to be the oldest objects in our Galaxy. The two chosen for the study, NGC288 and NGC362, have long been suspected to differ in age by 3 billion years. This inference is based on the colour-luminosity plot (Hertzsprung–Russell diagram) of the stars in the clusters: specifically, the point at which the stars depart from the main, hydrogen-burning sequence. In older clusters, the more massive, brighter stars will have exhausted their hydrogen cores and only the fainter, less massive stars will remain on the main sequence. The two clusters appear to have the same metallicity, a parameter which affects the turn-off luminosity at a given age and so might thereby introduce further uncertainty into any interpretation of difference in turn-off. This pair of clusters is therefore an ideal candidate for an age difference between two stellar systems that formed when the Galaxy first underwent collapse.

This originally appeared as "Slow Motion Galactic Birth," Joseph Silk, *Nature*, 351, 191, 16 May 1991.

However, previous studies made unwarranted assumptions about the metallicity of these clusters. Astronomers loosely group C, N and O, as well as Fe, under the rubric of 'metals', and there is reason to believe that in old stars in the galactic halo, the O/Fe abundance may differ from the solar value. Indeed at [Fe/H] <-1 (a compact notation that means Fe/H is less than 10 per cent of the solar value), [O/Fe] is about 0.5 (an enhancement of 3) relative to the solar value for field halo stars. A factor of 3 increase in O/Fe decreases the age by 2 Gyr from that derived for a given turn-off assuming the solar O/Fe ratio. Previous studies of globular clusters did not directly measure C, N, O and Fe in the same stars, but assumed a solar ratio of C, N, O to Fe. Moreover, the globular-cluster stars measured are stellar giants whose surface abundances may be contaminated by the nuclear reaction cycle involving C, N and O, which occurs in the stellar core. There is therefore serious cause for concern that the ages (and age differences if O/Fe varies from cluster to cluster) may be overestimated.

The Dickens study obtained accurate C, N, O and Fe abundances for samples of stars in the two globular clusters. The CNO cycle reactions can modify the abundances of C, N and O, but the sum C+N+O is conserved. The spectroscopic study reveals that although (C+N+O)/Fe is constant from star to star and from cluster to cluster, O varies markedly with, for example, C. The variations with C and with N in NGC362 confirm the trend expected for material that has undergone CNO cycling in stellar cores and then been mixed with surface matter. But NGC288 shows no evidence of such surface contamination in its giant stars. The two clusters are inferred to have had very different histories of mixing. The unmixed stars are enhanced in [O/Fe] by about 0.2, a slightly smaller enhancement than seen for halo-field dwarfs at the same metallicity.

The moral of this investigation to observers is: take heed—do not use O/Fe in a small sample of stars (an inevitable restriction because of the long integration times needed to acquire high-resolution spectra) to infer global trends of [(C+N+O)/H], this ratio being the key to age, or relative age, determinations. Lack of knowledge of [(C+N+O)/H] made previous attempts at age determinations uncertain by 3 Gyr or more.

Dickens et al. found that the two clusters have almost identical abundances of [(C+N+O)/H], as well as of [Fe/H], and conclude that the clusters therefore do differ in age, by 3 ± 1 billion years. The age difference comes tantalizingly close to showing that the Galaxy took longer than the free-fall time, about 1 billion years, to collapse. The globular clusters, because of their eccentric orbits, must have formed during this collapse. Gaseous dissipation of the energy released during contraction provides a means of slowing the collapse. Most models of galaxy formation require the round spheroid of the Galaxy to collapse dynamically on a free-fall time. A similar scheme applies to the formation of elliptical galaxies. The longer timescale, if confirmed, should inspire a new approach to galaxy formation theory.

Astronomers are left with another puzzle; the two globular clusters have very different horizontal branch (post-main sequence) morphologies: NGC362 has mostly red giants, whereas NGC288 has predominantly blue giants. A second parameter, other than age, is needed to explain this, and the new study shows that the [(C+N+O)/Fe] abundance is not the culprit. Theorists and observers alike have much work left to do before the formation of globular clusters is fully understood.

A Deep Look at Forming Galaxies

WITH AUGUST EVRARD

Young galaxies provide a test for theories of galaxy formation, a process that was the subject of a meeting[1]. Although no consensus emerged on when galaxies formed, the confrontation of new discoveries of very distant and possibly young galaxies with theoretical speculations about forming galaxies led to a stimulating exchange of ideas.

Protogalaxies have been the holy grail of observational astronomy for two decades, but the meeting was especially timely as it marked the discovery during the previous year of objects that were tantalizingly close to being confirmed as protogalaxies. The strongest claim was made by L. L. Cowie (Institute of Astronomy, Hawaii) and collaborators who had obtained optical and infrared CCD (charge-coupled device) images—down to 27 magnitude in the visible band, about as faint as previously obtained by A. Tyson (AT&T Bell Labs) in a similar deep survey of arcminute-square fields at high galactic latitude. The novel aspect of Cowie's work was that images in several colours revealed among the many faint galaxies a few which had flat spectral energy distributions.

Cowie and colleagues inferred from the energy distributions that these objects are forming galaxies undergoing an extreme burst of star formation at redshift $z = 2.8$–3.6, corresponding to perhaps 8–16×10^9 years ago. The total flux from these objects corresponds to a substantial fraction of that emitted by the early generations of stars that synthesized the met-

This originally appeared as "A Deep Look at Forming Galaxies," Joseph Silk and August Evrard, *Nature*, 335, 766–767, 27 October 1988.

als seen in galactic disks. It should be emphasized however that such large redshifts of ordinary galaxies are inferred and not directly measured.

Dramatic evidence for galaxy formation at high redshift comes from the optical identification of known luminous radio galaxies. A radio galaxy at $z = 3.395$ discovered by S. J. Lilly (Institute for Astronomy, Hawaii) seems from its visible and near-infrared colors to be forming new stars and to contain a substantial population of stars that formed at redshift 5 or before. This conclusion arises because, for the simplest models of galactic evolution to be valid, the infrared emission detected at a wavelength of 2 micrometres with this redshift must have been formed $1–2 \times 10^9$ years before $z = 3.395$. It should be noted that more recent (1992) observations show that the infrared flux was overestimated, thereby weakening the evidence for an older stellar population.

Lyman-α emission from this radio galaxy (H. Spinrad, University of California, Berkeley) has an elongated distribution extending over 100 kiloparsecs with a surface brightness resembling that of a giant disk. This object and at least seven other emission-line radio galaxies beyond $z = 2$ are elongated along the main axes of the associated radio-emitting regions. This contrasts with low-redshift galaxies whose radio lobes tend to lie along the minor axes of luminous elliptical galaxies (McCarthy, P. et al. Astrophys. J. 321, L39-L44; 1987). The implications of the apparently causal relations between these emissions remain to be explored.

The low intergalactic neutral hydrogen (HI) column density inferred from quasar absorption lines at $z \leq 4$—the 'Gunn-Peterson' constraint—also indicates star formation at high redshift. P. Shapiro (University of Texas) presented a model in which star formation starts at $z \geqslant 5$ that provides sufficient ionizing flux to satisfy the Gunn-Peterson limit. One consequence is that half the observed metal abundance of population I (young, metal-rich) stars must have been produced before $z = 3$. This must be reconciled with the interpretation of the objects at $z \approx 3$ discovered by Cowie and co-workers that can account for almost all of the early star formation in massive galaxies.

A. Wolfe (University of Pittsburgh) and collaborators have indirectly discovered protodisks at $z = 2–3$ by studying absorption spectra imposed on the emission of distant quasars. They interpret the absorption lines as broad (or 'damped') redshifted Lyman-a profiles appropriate to intervening disks of neutral hydrogen with column densities greater than 10^{20} cm^{-2}. The high-redshift systems occur so frequently that they cover about a fifth of the sky, and contain perhaps as much gas as is contained in stars within present galactic disks. In one case, illuminated by a quasar with extended radio emission, the neutral-hydrogen cloud has a radius of about 100 kiloparsecs, the putative scale of a protodisk. And the profiles of the 21-centimetre absorption lines in several cases have widths comparable to those seen in thin galactic disks.

Weak, narrow Lyman-α emission superimposed on a system of damped Lyman-α emission at $z = 2.5$ may have been observed (R. Hunstead and M. Pettini. Anglo-Australian Observatory). It seems that the rate of star formation is surprisingly low if this is a galactic disk that is predominantly gaseous and is yet to form most of its stars. If there are protodisks at $z = 2$–3, we would expect also to find protospheroids forming at higher redshift as spheroids have older stellar populations than disks.

A systematic search for optical counterparts to steep-spectrum (thus remote) faint radio galaxies (K. C. Chambers, Johns Hopkins University; G. K. Miley, Space Telescope Science Institute; W. J. M. Breughel, University of California, Berkeley) revealed six at $z > 2$. These include the current distance record holder at $z = 3.8$. The systems are dominated by line emission from hot gas, and the inferred rate of formation of massive stars required to excite the gas is comparable to that expected for a luminous protogalaxy. Alternatively, the gas may be ionized by radiation from the nucleus, although only improved observations will discriminate between these possibilities.

Assessing the relevance of radio galaxies to galaxy formation is made difficult by our lack of knowledge of how significant a proportion they are of all high-redshift galaxies. For example, cold-dark-matter theory predicts that most stars formed at low redshift, but permits some galaxies to form back to $z = 5$, and bare galactic nuclei even earlier. These predictions are sensitive to the degree of 'biasing' (the relative clustering strength of visible and dark matter) incorporated. Rival theories involving cosmic strings (R. Brandenburger, Brown University) or explosions triggered by the decay of superconducting cosmic strings (C. Thompson, Princeton University), are still too 'soft' to make testable predictions for observational astronomers.

The best prospect for resolving the uncertainties is the analysis of the deep optical redshift surveys being performed on square-degree fields by D. Koo (University of California, Santa Cruz) and R. Kron (Chicago University) and by R. Ellis and collaborators (Durham University). Preliminary results for these deep, optically unbiased, large-scale surveys, which include redshifts of several hundred objects down to 21 (and subsequently extended to 23) magnitude, show that extensive evolution must already be occurring by a redshift of about unity.

1. Titled *The Epoch of Galaxy Formation* (Durham, 18–20 July 1988).

In Search of Young Galaxies

The discovery of forming galaxies would impose important restrictions on cosmological theories and ultimately lead to a unique scenario for the evolution of the large-scale structure. Most observed galaxies are mature systems that are already well down the evolutionary scale, and the quasars discovered at record-breaking redshifts[1] are too distant to reveal any structure that could be associated with forming galaxies. Observations and theory suggest that galaxy formation occurred at the epoch $z \cong 5$ or earlier. (Redshift, z, measures the shift in the wavelength of spectral lines in a distant object relative to that of a nearby galaxy, caused by the expansion of the Universe, and is a useful measure of the time that light has taken to travel from its source[2].) However, N. Bergvall and S. Jörsäter present evidence that the nearby dwarf galaxy ESO400-G43 is newly formed.

It is natural to assume, from the evidence of our own mature Milky Way and other nearby galaxies, that young galaxies must be remote in time as well as in space—that is, galaxies are not forming now. This is the basis of one theory of galaxy formation, originally proposed by P.J.E. Peebles and R. Dicke to account for globular clusters[3]. This theory was resurrected[4] under the guise of a model in which galaxies formed from primordial isothermal fluctuations (that is, fluctuations in matter density but not in radiation density). Isothermal fluctuations allow small seeds, of mass similar to that of dwarf galaxies, to develop at early epochs and are consistent with an open Universe dominated by baryonic (ordinary) matter in which galaxy formation occurred at $z \cong 30$.

Young galaxies, however, are not necessarily present only in the very early history of the Universe. One theory that accords with this view and

This originally appeared as "In Search of Young Galaxies," Joseph Silk, *Nature*, 331, 561–562, 18 February 1988.

also appeals to hierarchical evolution of structure (in which dwarf galaxies form first, followed by larger and larger systems) but with nonlinear evolution occurring only relatively recently, has been successful in accounting for several aspects of the large-scale structure of the Universe. The idea that the Universe is dominated by cold dark matter comes from inflationary cosmology, which prescribes a Universe at critical density with adiabatic fluctuations. These fluctuations involve both matter and radiation and are constrained by inflationary cosmology to have considerable power on large scales. This means that in this model, large galaxies are still forming at the present epoch. As is usual in cosmology, one suspects that the truth lies somewhere in between.

A galaxy in process of formation can be characterized in several ways. Perhaps it has no old stars. It may have an exceptionally high rate of star formation that could not be sustained for long and that can account for the observed luminosity. Or it may be predominantly gaseous, the gas being destined to form stars. Perhaps the galaxy is still gas-rich but deficient in metals, a sure indicator of a primitive stage of chemical evolution. Or finally, the galaxy may be forming stars at high redshift, early in the evolution of the Universe. Candidate young galaxies generally satisfy at least two of these criteria.

Bergvall and Jörsäter describe[2] an object known as ESO400-G43 that fulfills several of these requirements. This blue, compact galaxy is forming stars prolifically, it is gas-rich, and it appears to contain no old stars. It also has a dark halo that Bergvall and Jörsäter traced by the 21-cm hydrogen radio emission and which extends to about 15 disk scale lengths. (This halo could be of use to cosmologists trying to derive mass limits for so-called heavy neutrinos. The inferred radius of the halo's core—beyond which the density tails off—and its inferred velocity dispersion give a lower limit of 80 eV for any single species of massive neutrino which would account for the dark-halo matter.) ESO400-G43 is also metal-poor, at about one-eighth of solar metallicity, and requires no old stars at all to account for the mass-to-light ratio of the disk.

This object should be warmly welcomed by advocates of the cold dark matter theory. A dark halo, an apparently young, chemically unevolved stellar component and a plentiful supply of gas are the ingredients expected for a young galaxy. Cold dark matter theory[5] prescribes the continuous rapid formation of many dwarf galaxies from the typical density fluctuations; larger fluctuations collapse later. A few of these dwarfs are incorporated into luminous galaxies. It is necessary to introduce the idea of biasing into this theory to suppress the formation of an excessive number of young galaxies. According to one version[6] of this idea, dwarf galaxy formation is interrupted by the onset of supernova-driven winds that strip out the gas. Perhaps during the fiery vigour of massive star formation, these objects would be recognizable as blue compact dwarfs. By the present epoch, however, one expects the Universe to be full of dwarf galaxies, pale remnants of blue, compact galaxies.

Occasionally, one might come across a rare, delayed fluctuation that recently collapsed. So Bergvall and Jörsäter[2] prefer to attribute the blue and compact nature of ESO400-G43 to such an event, so that no underlying component of older stars can be expected. An alternative explanation is that a dwarf galaxy has encountered a cloud of weakly enriched gas that fuels the formation of new stars as it falls inwards. The massive halo provides a gravitational well capable of accreting intergalactic gas. The density of such gas can be estimated from observations of enriched gas in rich clusters, and is probably adequate in less dense regions of the Universe to fuel accretion. Future observations, especially in the infrared, will refine the search for an older stellar population, and so help discriminate between the two options of virgin birth or resurrection for the origin of such remarkable blue compact dwarfs.

1. Shaver, P. Nature 330, 426 (1987).

2. Bergvall, N. & Jörsäter, S. Nature 331, 589–591 (1988).

3. Peebles, P.J.E. & Dicke, R. Astrophys. J. 154, 891–908 (1968).

4. Peebles, P.J.E. Astrophys. J. 315, L73-L76 (1987).

5. Bardeen, J., Bond, J.R., Kaiser, N. & Szalay, A. Astrophys. J. 304, 15–61 (1986).

6. Dekel, A. & Silk, J. Astrophys. J. 303, 39–55 (1986).

The First Stars

WITH BEATRIZ BARBUY AND ROGER CAYREL

We are made of carbon, nitrogen, oxygen, and traces of other heavy elements that are the ashes of burnt-out stars. Around us we see dense clouds of molecular gas and dust. Embedded deep within these clouds are the embryos of newly forming stars. From this we infer that our stellar predecessors, ancestors of the sun, were themselves the remnants of a previous generation of stars. Debris from these long-dead stars enriched the molecular gas clouds in heavy elements. Each successive generation of new stars carried within it the memory of earlier generations, fossilized in the form of heavy elements.

But where did the first stars come from, and where are they today?

The First Billion Years

The Universe itself began in a fiery explosion that we call the Big Bang. The density and temperature greatly exceeded that within any known stars. The matter of the Universe consisted initially of protons, electrons and neutrons, together with neutrinos and radiation. The radiation, or at least its greatly expanded and cooled remnant, is seen today as the cosmic microwave blackbody radiation: this echo of the primordial fireball constitutes the single most persuasive piece of evidence for the Big-Bang theory. Today, the temperature of the radiation was about three degrees Kelvin, and most of the energy is in microwave or radio photons. But about

Published originally as "Les Premieres Etoiles," Beatriz Barbuy, Roger Cayrel and Joseph Silk, *La Recherche*, 9(180). 1060–1069, September 1986.

one or two minutes after the explosion, the temperature of the Universe was one billion degrees. The radiation was in the form of energetic gamma rays.

The particles present, protons, neutrons, and electrons, moved around with considerable speed, about one percent of that of light, and this was sufficient for them to undergo nuclear reactions. At higher temperatures, any synthesized elements would be destroyed by the gamma rays, but at a billion degrees or so, one begins to form light elements, most notably deuterium and helium. Mostly, one makes helium: about 25 percent by mass ends up as helium nuclei, and only about one-hundredth of a percent or less as deuterium. These light elements survive today, and are seen in the spectra of stars and interstellar gas clouds. Indeed, astronomers observe that approximately 25 percent by mass of stars, interstellar clouds, and galaxies, at least for those objects with measurable helium lines, is in the form of helium. This result holds despite the known fact that stars augment the initial helium supply: they produce helium as they burn hydrogen for fuel in their cores. Even the most primitive objects known, that is, the most metal-poor systems, contain more or less this universal abundance of helium, thereby providing observational verification of the Big Bang prediction of the cosmic helium abundance.

The deuterium observed in interstellar clouds and in the solar system is also believed to have been synthesized in the Big Bang, primarily for the reason that astronomers have had great difficulty in hypothesizing any other origin for the deuterium. It is a fragile element, easily destroyed in stars, and only a small amount, about 0.003 percent by mass, is observed. The abundance of deuterium is extremely sensitive to the baryonic mass fraction of the Universe. With a high baryon fraction, the deuterium is largely destroyed, to be incorporated into helium. At low baryon fraction, slightly less helium is made, but the deuterium abundance may be orders of magnitude larger. Comparison of the predicted and observed abundances of the light elements constrains the baryon fraction in the Universe to be between 2 and 6 percent of the critical density for closing the Universe. Since we observe about 1 percent in the form of luminous matter (baryons), the dominant mass in baryons is observed to be dark. This could be in the form of compact remnants of stars or objects too low in mass (below 0.08 the mass of the sun) to have ignited their hydrogen and become stars.

But apart from slight traces of lithium, beryllium, and boron, the Big Bang failed to produce any heavier elements. Helium has atomic mass 4, consisting of 2 protons and 2 neutrons: there is no stable element of mass 5 or of 8. Hence the link in the chain was broken: as the Universe cooled, there was not time enough to make any carbon, of mass 12, or any heavier elements. In this way George Gamow's glorious vision of the origin of all the known elements in the hot Big Bang failed.

A great step forward was taken by the English astrophysicist Fred Hoyle. He realized that in the hot cores of stars, after exhaustion of their hydrogen fuel, there was the possibility of fusing three helium nuclei directly into a carbon nucleus. In this way, the bottleneck arising from the absence of stable elements at masses 5 and 8 could be circumvented. But this could only occur inside a star, where one had both high temperatures and densities, and time enough for carbon synthesis to occur. The fuel supply in the stellar core is gradually exhausted as elements as heavy as iron are produced. Each stage of nucleosynthesis liberates thermonuclear energy that helps support the dying stars. Finally, in a massive star, only iron remains in the core. Iron is the most stable of elements, and releases no additional energy by fusion: rather, energy must be supplied to fuse still heavier elements. When the core fuel supply is exhausted, the core collapses to form a compact neutron star, and the release of energy expels the outer envelope of the star in a gigantic explosion. These explosions are seen as supernovae, and elements heavier than iron are produced via the neutron irradiation of ejecta that occurs during the explosion.

Many neutron stars have been discovered: they are found as pulsars. Rapid pulses of radio emission, typically every second, occur with unprecedented regularity from pulsars. These prove that the emitting object must be spinning, emitting a beam of radiation, and have a radius of only about 10 kilometers, yet have a mass equal to that of the sun. The density of matter inside a neutron star is equal to that inside an elementary particle: protons and electrons have been squeezed together to form a single gigantic nucleus. One teaspoonful of neutron star matter weights about ten billion tons.

Supernova explosions are the fate of stars more massive than about 10 times the mass of the sun. Stars of a lower mass undergo a more quiescent fate. After the hydrogen supply is exhausted in the stellar core, they swell up into red giants and supergiants, and then expel shells of matter enriched in nitrogen and carbon. Finally, the core contracts to form a white dwarf star. This is a compact state of matter in which the star is supported not by ordinary pressure of a hot gas, but by quantum pressure. In this bizarre state, the uncertainty in electron positions predicted by Heisenberg's uncertainty principle contributes to an effective pressure that can balance the attractive pull of gravity. Many white dwarfs have been observed by astronomers: the typical size is that of the earth, yet the mass is that of the sun. One teaspoonful of matter in a white dwarf weighs about a thousand tons.

How Did the First Stars Form?

About one billion years after the Big Bang, the first primordial gas clouds condensed out of the expanding Universe. Infinitesimal density fluctuations had gradually amplified by means of their own self-gravity, eventually accumulating sufficient density contrast to pull away from the general expansion and contract. Opinions differ as to the precise mass of these pregalactic clouds, but most theories of galaxy formation converge on a mass of about 10^5 or 10^6 solar masses. This is a mass scale that actually is predicted by the Big-Bang theory itself, since it is the minimum mass cloud that can develop from fluctuations in the early Universe. A galaxy forms by the clustering and merging together of many thousands or hundreds of thousands of such clouds. According to P. James Peebles and Robert Dicke of Princeton University, the surviving clouds are to be identified with the 149 globular star clusters that contain the oldest stars in our galaxy.

A gas cloud has little choice but to collapse and fragment into stars. Only the random motions of its atoms provide a pressure that resists gravity, but only for a brief time: the atoms collide, radiate, and lose their kinetic energy of motion. Interstellar gas clouds radiate because of the presence of heavy atoms such as carbon and iron. Collisions of hydrogen atoms with these species excite the electrons around the heavy element nuclei: an instant later, the heavy atoms radiate away their excess energy and return to their unexcited ground state. In this way, energy of atomic random motions, that determines the pressure force, is lost from the cloud. In a primordial cloud, there is no iron or carbon, and one might at first think that since cooling by excitation of heavy atoms does not occur, the cloud is destined not to cool and continue its collapse. In fact, a small amount of molecular hydrogen forms. There are a few electrons left over from the early ionized state of the cloud in the hot Big Bang: these electrons combine with hydrogen atoms to form negative hydrogen ions, which in turn capture protons to form hydrogen molecules. Now a hydrogen molecule has many low-lying energy levels associated with the rotation and vibration of the molecular system, and cooling can consequently continue in the primordial cloud. The outward pressure force decreases, and the cloud inexorably collapses. Only if the collapse were very regular and spherically symmetric could it conceivably lead to the formation of a single supermassive star. This may happen rarely, and such objects could have unique consequences, for they would continue to collapse to form a million-solar-mass black hole. Supermassive black holes are believed to exist: there almost certainly is one at the centre of our galaxy. But they are probably rather rare objects. A much more likely sequence of events is as follows. The collapse will be very asymmetric, and lead to formation of dense sheets and filaments. These are unstable: gravitational forces cause them to break up on a scale corresponding to their thickness into many small fragments. These fragments are the precursors of individual stars. The million-solar-mass gas cloud has formed a cluster of stars.

The principal difference between a primordial cloud and a molecular cloud in the present interstellar medium is that the latter is highly enriched in heavy elements. Detailed calculations show that because of the formation of molecular hydrogen in the primordial clouds, fragmentation in both cases can proceed down to very small masses, of order one-tenth or one-hundredth that of the sun. This suggests that one forms a wide range of stellar masses, since the fragmentation theory only sets a lower bound on fragment masses. Processes such as coalescence between fragments guarantee that some massive stars will also form. In the present interstellar medium, one finds stars ranging between one-tenth and one hundred solar masses. Stars of much lower mass would not ignite their nuclear fuel and hence be invisible, whereas more massive stars would be unstable and may not even be able to form.

There are some differences, however, between primordial and present-day star formation. Perhaps the most significant is that once even a single massive star has formed in a primordial cloud, its radiation field is capable of dissociating all of the molecular hydrogen. The ultraviolet radiation field will prevent molecular hydrogen from reforming, and the only resort remaining for the primordial cloud will be to cool via collisions between atoms of atomic hydrogen and radiation of Lyman alpha photons. These photons of energy 13.6 electron volts or wavelength 1216 Angstroms are produced whenever a hydrogen atom is excited. However, it takes considerably more energy to excite the electron of a hydrogen atom than to set a molecule into rotation. Instead of cooling at one thousand degrees Kelvin, in the absence of any molecular or heavy elements, the cloud, as it collapses, must heat up to about ten thousand degrees Kelvin. At this temperature, continued collapse can occur as the cloud radiates its internal energy away via Lyman alpha photons. The net result, however, is that at a specified density the fragments which form are now much more massive than before, primarily because of the elevated temperature. However, the opacity is another more complicated ingredient that helps determine whether fragments can undergo further subfragmentation as collapse continues. The opacity is lower when heavy elements are absent, hence collapse can indeed continue further.

The crucial issue is survival to form a distinct star: fragments may not survive as distinct entities but merge together or else grow by accreting ambient gas. At the warmer temperature of a primordial cloud, accretion is inevitably more important, as the sound speed is larger and subsonic turbulence is therefore more vigorous. A modern view of star formation regards the entire process as being an incoherent, inhomogeneous mix of clumps that formed by fragmentation, and then grew by accreting gas from their surroundings. There is even circumstantial evidence from observations of star-forming regions that when gravitational interactions stir up a dense interstellar cloud, the formation of progressively more massive stars is formed. Such interactions occur in spiral arms, and may accompany particularly vigorous star formation, where there is considerable energy input associated with the outpouring of radiation and wind from massive stars, and with the explosions of dying stars. Large-scale gravitational tidal forces, associated with close interactions of galaxies, are responsible for the most violent episodes of star formation, or starbursts, when much of the interstellar medium is triggered into forming stars on a relatively rapid time-scale compared to that of the Milky Way galaxy.

In the early Universe, galaxy formation is itself regarded as a continual process of gaseous fragments merging to form larger fragments and eventually galaxies. Perhaps a forming galaxy is a sequence of starbursts as massive gas clouds merge together, forming stars at a greatly elevated rate compared to that viewed at the present epoch. Such an enhanced rate is required for primordial star formation, in order to get the heavy elements in place sufficiently early. Thus the consensus view might be that primordial clouds should form stars which span a mass range between 0.1 solar mass and 100 solar masses, but with a very considerable enhancement, compared to present-day star formation, in the numbers of more massive stars.

What could be left over from the very first burst of star formation? We now turn our attention to two observable tests. One is the search for stars with extremely low metallicities. Any star with a metallicity below, say, one-thousandth that of the level observed for the sun is likely to be from this initial burst, since theoretical arguments suggest that heavy element cooling only begins to dominate at higher metallicity. A second is to look for a signature of the precursor population in terms of abundance and isotope ratios in extreme population II stars. The results of such searches have hitherto been inconclusive. If primordial stars exist, as they would if their masses spanned the range below that of the sun, then they must be exceedingly rare. We know that massive primordial stars were present from their pollution: the heavy elements we observe in the oldest extinct stars were produced in supernova explosions some 10 or 12 billion years ago. An indication that the precursor stars weighted some 20 or 30 solar masses comes from the ratio of the abundance of oxygen to that of iron seen in intergalactic gas in two nearby galaxy clusters. The gas is enriched in iron to half the abundance of iron in the sun. The high abundance of intergalactic iron requires these stars to have formed in great numbers near the beginning of the galaxy, where many massive stars would have died in fiery supernova explosions and released enriched debris into the interstellar gas. The enhanced oxygen points directly to the massive star precursors: stellar evolution theory fingers such stars as the unique hosts of the requisite overabundance of oxygen.

The paucity of extremely metal-poor stars has been explained by two alternative hypotheses. Perhaps the early stars were predominantly, although not exclusively, more massive than the sun. Very few survive today in the halo to bear witness to the past. This hypothesis has a radical consequence: the remnants of these stars, white dwarfs, neutron stars, and even black holes, must still populate the halo. Conceivably, they could exist in large enough numbers to account for the entire "dark mass" of the halo, some 10 times more mass being inferred from the galactic rotation velocity than is directly seen in stars.

There is a more conservative viewpoint. Perhaps the first generation of stars polluted its environment with great efficiency. The next generation of stars to form would thus contain some heavy elements, and successive generations even more so. One might then naturally expect there to be very few primordial present survivors from the beginning of the galaxy.

A consequence of the conservative view is that star clusters "self-pollute." The natural means to disperse the heavy elements by supernova ejecta can only be achieved on the scale of clouds that eventually form globular clusters. Thus, as one compares one globular star cluster to another, these star clusters containing the oldest stars in the galaxy, one would expect to find a broad dispersion in metallicity. Each cluster must have experienced a somewhat different history of star formation and enrichment. There is some evidence of such a spread from cluster to cluster, although the fact that within a given cluster, the stars are remarkably homogeneous in composition is difficult to reconcile with self-pollution. It is as though the globular clusters formed from diverse regions of a much larger gas cloud.

The radical hypothesis provides a more natural means of seeding the entire protogalaxy with heavy elements before the globular clusters formed. It has led to even more radical predictions. The dark halo could consist of white dwarfs. Occasional mergers of close pairs of binary white dwarfs would form neutron stars that might be observable in the remote and otherwise dark regions of the halo as radio pulsars or even as gamma ray bursters. The predominance of massive stars in the protogalaxy may represent a mild version of a vigorous phenomenon that appears to occur in "starbursts." These are galaxies that undergo an episode of greatly enhanced star formation induced by a close encounter or merger with another galaxy. The theory of galaxy formation tells us that a galaxy forms out of the gravitational aggregation and coalescence of many smaller gas clouds. Perhaps our Milky Way "protogalaxy" underwent a starburst, or even a series of starbursts. This dramatic firework display would have settled down after a few hundred billion years, as the gas supply was exhausted, to form first the halo, and eventually the disk of the galaxy that we recognize today as the Milky Way.

The first stars are evidently elusive. If we fail to find them in the halo of our galaxy, and that means essentially within a few thousand parsecs of the sun, we may yet uncover their counterparts in starbursting galaxies visible at very great distances. We look back in time as we sample distant parts of the Universe with the largest telescopes, and one can study remote galaxies at a look-backtime of ten or even fifteen billion years, to an epoch when the Milky Way was barely a glimmer. Direct study of galaxy formation will ultimately be accessible to astronomers, although, as with the first stars, discovery of a galaxy undergoing the birth process has hitherto proved elusive. However, forming galaxies must have been luminous, for they contained the massive stars that made the elements in the oldest stars. Success is guaranteed, eventually, although the optimal choice of wavelength is but one of the obstacles to be surmounted. Such forming galaxies may be surrounded by cocoons of dust and be visible in the far infrared part of the spectrum that is inaccessible from the ground. Future telescopes in space may be the crucial stepping stones to the first stars.

Did the Tail Wag the Cosmic Dog?

One of the problems that preoccupies cosmologists concerns the distribution of matter over the large scales of size we see in the Universe today. Any theory of the origin and evolution of the Universe should incorporate processes leading to the hierarchy of galaxies and clusters of galaxies that have formed in the course of cosmic time. Within the Big Bang theory that is central to most cosmologists' way of thinking, neither of the two leading theories of galaxy formation from fluctuations in the primordial matter seems fully to satisfy the observations. A new theory proposes that primeval fluctuations on a much smaller scale than galaxies may provoke the growth of much larger-scale fluctuations. The formation of such seeds could provide the key to understanding galaxy formation.

Understanding the large-scale structure of the Universe has proved surprisingly difficult. We see galaxies, typically of 10^{10}–10^{11} solar masses, groups of galaxies, and great clusters and superclusters of galaxies, on scales of up to some tens of megaparsecs, as well as many dwarf galaxies, extending in mass down to 10^6 solar masses, but which can only be seen easily in the vicinity of our own Galaxy. This diverse array of structures is what cosmologists have tried to account for through the evolution of density fluctuations in an initially homogeneous, or nearly so, Big Bang. The theory has met with mixed success. Two natural scales emerge, but neither corresponds to that of a galaxy like our own Milky Way galaxy.

The most popular version starts with adiabatic fluctuations—fluctuations whose entropy per unit mass is everywhere the same and equal to

Originally published as "Did the Tail Wag the Cosmic Dog?" Joseph Silk, *Nature*, 303, 200–201, 19 May 1983.

that of the Universe itself. Theories of particle creation in the initial instants of the Big Bang result in a universal value of the entropy per baryon, which otherwise is one of the unexplained mysteries of the cosmos. The final scale that emerges from an initial, more or less arbitrary, spectrum of adiabatic density fluctuations is that of a large galaxy or a small cluster of galaxies. All primordial structure on smaller scales was erased by the diffusive effects of the radiation in the early Universe.

The ensuing implications for galaxy formation lead to a rather convoluted series of events. The large-scale inhomogeneities from which small-scale structure, once present, has been erased by well-understood physical processes, eventually collapse under the influence of gravity. These gigantic clouds fragment by way of a pancake-like collapse to compressed sheets of matter. The initial fragments are clouds of masses comparable with those of dwarf galaxies, and these clouds rapidly coagulate to form galaxies of larger mass.

A variant on the adiabatic theory begins with isothermal fluctuations. These are entropy variations, and are not in accordance with theories of baryosynthesis and a universal entropy per baryon. However, isothermal fluctuations could also be created by subsequent events, involving energy dissipation, for example, and their evolution does lead to the survival of primordial small-scale structure. The characteristic scale at the onset of galaxy formation corresponds to between one-tenth and one million solar masses. Above this scale, gravity dominates pressure forces once gravitational collapse can occur after the epoch of matter-radiation decoupling at a million years or so after the Big Bang.

Still smaller scales are possible if cold dark matter such as axions dominate the mass-density of the Universe. These hypothetical elementary particles, predicted by a certain theory of nuclear interactions, are bosons—whose phase space density, unlike that of neutrinos, is not restricted by the Pauli exclusion principle. Hence axions have become an interesting but hypothetical source of dark matter, since they could be dynamically important over a very wide range of scales. From the perspective of primordial fluctuations, the fact that axions are collisionless particles over much of the evolution of the Universe means that preexisting fluctuations would not have been erased down to scales of over a solar mass.

None of this is very promising for the problem of understanding how a typical luminous galaxy has formed. If the answer is indeed that galaxies have formed by complex dissipational and dynamical processes, and any memory of primordial structure has been erased, this would inevitably disappoint many cosmologists—not all, however, as is illustrated by an interesting paper by Craig Hogan.[1]

Hogan postulates the presence of initial seeds, perhaps of stellar mass, and shows that a not-unnatural sequence of events in the early Universe may lead to the formation of secondary structures on scales up to those of the great galaxy clusters. The only truly primordial fluctuations are the stellar seeds. The release of large amounts of ionizing radiation from the seeds, which may be either massive stars or supermassive quasar-like compact objects, heats and ionizes a large volume of gas. This has the effect of generating a gradient of pressure that in turn drives large density fluctuations. Thus, small-scale seeds have stimulated formation of large-scale irregularities. In time, the inhomogeneities grow larger by gravitational instability and eventually form bound structures.

Two attractive features emerge naturally from Hogan's scenario. The limiting scale on which fluctuations are generated is the maximum Jeans mass in the history of the Universe, for the Jeans scale determines the maximum scale over which pressure-driven fluctuations can propagate. This scale is about 10^{16} solar masses in the conventional hot Big Bang theory. Furthermore, if the seeds form by a random process, the fluctuation distribution will then be spatially uncorrelated. Now the gravitational clustering theory for formation of galaxy groups and clusters has been shown to give realistic models of the observed galaxy distribution. Implications of this theory are that the initial fluctuations were more or less random, and that the largest scales to have collapsed were of about 10^{15} or 10^{16} solar masses.

All of this agrees nicely with the theory of primordial seeds—but can the theory be tested? One implication is that there should not be any intrinsic large-scale structure in the Universe. Preliminary reports of the detection of an intrinsic quadrupole anisotropy in the cosmic background radiation appeared fatal to this theory. However, between the submission of Hogan's paper and its publication, the quadrupole anisotropy disappeared: that is to say, more careful experiments failed to confirm it. These experiments set a low upper limit of about 3 parts in 10^5 on any possible quadrupole anisotropy.[2]

Forthcoming observations of anisotropy on intermediate angular scales of a few degrees should eventually provide a sensitive test of the primordial seed theory. For if intrinsic primordial structure were present on galactic cluster scales, fluctuations in the background radiation are inevitable at a level of about one part in 10^5.

The ultimate question in such a theory, of course, concerns the nature of Hogan's seeds. Here one runs into a fundamental difficulty. The stability of the hot Big Bang model to the growth of infinitesimal perturbations has been thoroughly investigated. Certainly, at epochs where the physics is well understood, it seems to be stable. At these epochs, the horizon size, which determines the maximum scale over which causally generated fluctuations could spontaneously arise in a phase transition, contains far less than a solar mass. Hence it seems impossible to generate the required stellar mass seeds.

There are two ways to circumvent this conclusion. At much earlier epochs, close to that of the Big Bang singularity itself, the horizon structure of the Universe may differ greatly from that predicted in the conventional Big Bang model. Such is the case if the Universe undergoes a prolonged phase of exponential inflation, as is expected in certain theories of grand unification in particle physics. Fluctuations may rise spontaneously during the transition from an inflating universe to the conventionally expanding model, although there seems no reason to expect their scales to be limited in any way. More exotic types of phase transition have also been postulated, involving strings or walls of vacuum energy that, before eventually decaying, can generate inhomogeneities.

An alternative and even more radical approach is to dispense with the hot Big Bang entirely. It is the high temperature and entropy of this model that creates such difficulties for fluctuation growth. In an initially cold (or even lukewarm) Big Bang, spontaneous growth on stellar mass-scales appears likely, if not inevitable, during the early transition from a solid to a gaseous phase. This model may provide the required seeds, but of course it creates other difficulties, notably the need for alternative explanations of the cosmic background radiation and of the light-element abundances, two pillars of the conventional Big Bang theory.

Sadly for Hogan's hypothesis, the COBE satellite reported in 1992 the detection of a quadrupole anisotropy in the cosmic microwave background, at a level of about 6 parts in 10^6. Unless there is some more local explanation of this very large-scale anisotropy, a possibility which seems contrived and unlikely, it would seem that the primordial seed hypothesis can be discarded.

1.Monthly Notices of the Royal Astronomical Society 202, 1101; 1983.

2.Nature 302, 478; 1983.

Primordial Helium Abundance and Big-Bang Cosmology

The primordial abundance of helium is one of the three great predictions of the Big-Bang theory. The theory holds that helium was synthesized during the first minutes of the expansion, and predicts its mass fraction to be about 25 per cent of the ordinary matter in the Universe. Of course, stars are also gigantic fusion reactors that convert hydrogen into helium, so it is crucial to disentangle the stellar contribution to the observed helium abundance from its primordial cosmic value. A workshop devoted to primordial helium was convened in February 1983[1] by the European Southern Observatory at Garching and provided a unique opportunity to assess the current status of theoretical predictions and observations.

Over the previous four years, cosmologists had honed predictions of the helium mass fraction (Y) to three significant figures. A higher-density universe synthesizes helium more efficiently, increasing Y slightly. G. Steigman (Bartol Foundation, University of Delaware) reported on the role of different types of neutrinos during nucleosynthesis. Physicists have discovered three neutrino species: the electron, the muon and τ neutrinos; and in principle, further species await discovery. Each additional species contributes to the energy density of the Universe, and thereby slightly speeds up the expansion during the high-temperature regime when nucleosynthesis occurs. As helium synthesis competes primarily with the decay of free neutrons—a neutron possessing a half life of about 10.5 min—the speed-up means that neutron capture by protons occurs more efficiently and produces slightly more helium.

Originally published as "Primordial Helium Abundance and Big-Bang Cosmology," Joseph Silk, *Nature*, 302, 382–383, 31 March 1983.

Stars also produce and eject helium into the interstellar medium. Allowance must be made for this enrichment, and the primordial value of the helium abundance is denoted by Y_p. With three neutrino species, the predicted Y_p is 0.25 for a universe that is at one-tenth of the critical density for closure. In a closed universe, Y_p increases to 0.26. Only if the baryonic density is very low indeed, corresponding to a density of 1 per cent of the closure value, is Y_p reduced to 0.22. The presence of further, as yet undiscovered, neutrino species has the effect of increasing Y_p. Determination of Y_p therefore has potential implications of immense import for cosmology and particle physics.

Observers had a field day at Garching. Total failure to agree on a unique value for Y_p left them concluding that any value between 0.22 and 0.26 was consistent with their data. The problems began surprisingly close to home, with the Sun. D. Gough (University of Cambridge) reviewed the remarkable results of solar seismology. Observations of solar oscillations probe the structure of the Sun. Surface modes propagate in the convection zone, the depth of which depends on the solar helium abundance. If the helium abundance is increased, the central temperature rises to maintain the same central pressure and results in a steeper temperature gradient and convection zone. The observed oscillations can be fit with $Y = 0.25$ (± 0.01), the principal uncertainty arising from the equation of state of the Sun. However, the inferred central temperature is high enough to produce a solar neutrino flux that is well above experimental limits, which require $Y = 0.19$. The resolution may well be that the neutrino possesses a slight mass. Even one-millionth of an electron volt would suffice to allow electron neutrinos to oscillate with muon and τ neurinos between the Sun and the Earth, thereby removing the experimental discrepancy. Unfortunately, current experimental limits on the neutrino mass are only of order 1 electron volt.

Going outside the Solar System, M. Peimbert (Instituto de Astronomia, Mexico City) reported on the helium abundances in planetary nebulae, while P. Mezger (Max-Planck-Institut für Radioastronomie, Bonn) and P. Shaver (Garching) discussed Y_p determinations in HII regions. As HII regions are young (typically 1 Myr) and are expected to have acquired some helium from stellar enrichment of the interstellar gas, emission from and singly ionized helium is seen, and the principal uncertainty in evaluating Y is in estimating the amount of neutral helium. A well studied HII region, such as the Orion nebula, has $Y = 0.29$, indicating very considerable enrichment of helium since the Solar System formed. However, attempts to infer the primordial helium abundance are much more controversial. By observing planetary nebulae and HII regions that display a range in metallicity, Z, Peimbert found a correlation between part of the helium abundance, ΔY, and the metallicity variations, ΔZ.

Attributing ΔY to the stellar enrichment that produced the observed heavy elements, the primordial helium abundance can then be inferred. Peimbert concluded that $Y_p = 0.22 \pm 0.01$, but this low value depended critically on his finding that $\Delta Y/\Delta Z = 3$.

The danger in this procedure was demonstrated by D. Kunth (Institut d'Astrophysique, Paris), who studied several extragalactic HII regions. These were chosen because of their low metallicity, leading one to expect that they should be less enriched by stellar nucleosynthesis than galactic HII regions. Kunth found no evidence for any correlation between helium abundance and metallicity. His conclusion was that $Y_p = 0.25$ (± 0.003) in the regions he studied, unless one regards this value as an upper limit, since he could not rule out any stellar contribution to the observed helium.

With observers failing to establish definitively any $\Delta Y/\Delta Z$ correlation, crucial for interpolating to primordial Y, the theoreticians attempted to resolve the problem. It soon became apparent that theoretical models of the evolution of massive stars—prime candidates for substantial helium production—are in a similar state of disarray. The culprit here is the uncertain role of mass loss during stellar evolution. A. Maeder (Geneva Observatory) concluded that practically any correlation with $0 < \Delta Y/\Delta Z \leq 3$ was possible, when stellar models were combined with simple models of galactic evolution.

At this stage, the consensus on primordial helium was that with any value between 0.22 and 0.25 (or even 0.26) for Y_p being allowed by observation, practically any Big-Bang cosmology was possible. A very open or a closed universe was equally possible. However, helium is not the only element produced in the Big Bang; other important elements are deuterium and lithium. Observations of the abundance of these elements were reported that promised to have a resounding impact on our understanding of galactic evolution, and indirectly constrain the amount of helium enrichment that is so crucial to determining Y_p.

Deuterium is partially destroyed by stars. The Big-Bang nucleosynthesis calculations predict a primordial deuterium abundance that is extremely sensitive to the mean baryonic density of the Universe. In a high-density or nearly closed universe, little deuterium survives the first few minutes. In addition, the stellar destruction (or astration) factor is unknown, and must be inferred from observations. The interstellar medium provides an environment where the current deuterium abundance can be measured. C. Laurent (Verrieres-le-Buisson) reported a re-analysis of deuterium absorption-line observations in galactic HI regions. One of the difficulties which plagued earlier analysis of these UV data obtained with the Copernicus satellite was the apparent variations along different lines of sight of the measured deuterium abundances relative to hydrogen. Laurent showed that small amounts of circumstellar hydrogen moving at high velocity were contributing to the apparent deuterium line profiles. In one instance, time variations of the line profiles confirmed this interpretation. Correction for this effect yielded an interstellar deuterium abundance of 5 (±3) $\times 10^{-6}$ by number relative to hydrogen.

D. Gautier (Observatoire de Paris, Meudon) presented results from the Voyager space probe studies of Jupiter and Saturn. Deuterated molecules were observed and, unlike the situation in interstellar molecular clouds, the temperature is believed to be too low for any appreciable fractionation to have occurred. The inferred presolar abundance of deuterium is (2 – 5) $\times 10^{-5}$ deuterium atoms per hydrogen atom. Since some five billion years of galactic evolution have elapsed between the formation of Jupiter and observed interstellar clouds, one infers an astration factor over this period of about 2, and perhaps as large as 8.

Such considerable astration exceeds that predicted in models of galactic evolution. Further confirmation that drastic revision of our understanding of galactic evolution is needed was provided by two observational coups from France. F. Spite (Observatoire de Paris, Meudon) presented his detections of lithium in old halo dwarf stars. The observed abundance of 10^{-10} lithium atoms per hydrogen atom does not depend on stellar temperature or metallicity, and so presumably reflects a more or less primordial value. As stressed by J. Audouze (Institut d'Astrophysique) in his introductory survey, such a low primordial abundance of lithium is practically impossible to reconcile with the revised interstellar deuterium abundance, unless very considerable galactic astration of deuterium is assumed. R. Ferlet (Verrieres-le-Buisson) announced the first detection of the lithium isotope ratio in the interstellar medium. Along one line of sight, he found that $^{7}Li/^{6}Li \approx 40$. Comparison with the Solar System value of 12.5 indicates that, since only ^{7}Li is produced in substantial amounts by the Big Bang, considerable stellar destruction or possibly synthesis of lithium must have occurred since formation of the Solar System. Finally, the interstellar abundance of the isotope ^{3}He was reported by T. Wilson (Max-Planck-Institut für Radioastronomie, Bonn), who detected the spin-flip radiofrequency transition of $^{3}He^{+}$ from three galactic HII regions and was able to measure directly the column density of $^{3}He^{+}$. He found that $^{3}He^{+}/H^{+} = 5 \times 10^{-5} - 10^{-4}$ to within a factor of two for the ionized gas that he studied. The ionization correction to infer ^{3}He is unlikely to be large, and one may compare the result with the Solar System meteoritic abundance $^3He/H = 2 \times 10^{-5}$. Since ^3He is believed to be produced primarily by astration of deuterium, this result is again indicative of large deuterium astration over the past 5×10^9 yr.

The general consensus among astronomers was that the standard model of Big-Bang nucleosynthesis faced no insuperable difficulties. Were new determinations of Y_p ultimately to restrict the model to the point where it was incompatible with the independently derived baryonic density, for example, there always remains the possibility of synthesizing much, if not all, of primordial helium in pregalactic stars. In an initially cold or tepid Big Bang, or in a highly anisotropic initial Big Bang, no helium need be produced. Pregalactic synthesis of helium by massive stars may also lead to a model which can naturally account for the other light elements. J. Audouze and J. Silk (Institut d'Astrophysique, CNRS, Paris) noted that in the absence of any pre-existing helium, cosmic-ray helium nuclei accelerated by winds from massive initially pure hydrogen stars would directly produce ample deuterium by spallation on ambient hydrogen without overproducing lithium or other heavy elements.

The principle excitement, however, lay not in such heretical speculations, but rather in the growing realization that galactic evolution models need severe revision. The apparent increases in abundance of both helium isotopes, as well as the decreases in deuterium and ^6Li, since Solar System formation all point to very considerable astration over the past five billion years. The challenge is to theoreticians to develop models of galactic evolution that are capable of satisfying these constraints. As for observers, the value of the primordial helium abundance remains as elusive as ever, and much work has still to be done to provide values of Y_p that are capable of effectively constraining Big-Bang cosmology.

Despite the decade since these results were reported, subsequent developments have been modest. Corrections to dwarf galaxy abundance determinations have been refined, and the preferred value of Y_p is 0.24. The correction from the observed Y to the primordial value, Y_p, involves a correction for stellar production of He that Maeder now finds to be as large as $\frac{\Delta Y}{\Delta Z} = 6$. LEP has measured 3 neutrino species, bringing prediction into agreement with observation if the remaining parameter, baryon density, satisfies

$$\Omega_b h^2 = 0.015(\pm 0.005).$$

Other highlights include measurement of the ^6Li isotope in a hot star $(\frac{^7Li}{^6Li} \approx 20)$, as well as improved errors on its interstellar abundance, detection of deuterium towards a nearby star ($\frac{^2H}{H} \approx 2 \times 10^{-5}$), and measurement of beryllium and boron in halo stars at levels that suggest a spallation origin.

---

1. ESO Workshop on Primordial Helium, 2–3 February 1983, Garching bei Munchen.

Historic Fluctuations

Cosmologists generally believe that galaxies formed as a consequence of the growth of tiny density fluctuations in the very early Universe. The remarkable uniformity of the cosmic microwave background radiation has provided convincing evidence for the high degree of regularity of the Universe at an early epoch but in the early 1980's two groups of astronomers, at Florence and Princeton, reported evidence for a very weak anisotropy on an angular scale of 90°, which is probably produced by similar matter fluctuations to those from which the galaxies evolved. These early results were subsequently retracted, but in 1992 the COBE satellite provided definitive evidence of fluctuations over 10° to 90° at a level of 1 part in 10^5.

One of the most tantalizing issues in cosmology concerns the origin of these fluctuations. It is thought that the initial conditions of the big bang might provide an answer. Perhaps the Universe is the way it is because of the way it was. Surely the resourceful cosmologist can improve on this hollow echo of creationism, for if the initial conditions were very different, then we would not be here to take stock of the situation. This concept of the observer's role in determining the state of the Universe has been elevated by Robert Dicke and Brandon Carter to the status of a fundamental cosmological principle: the anthropic principle. Its power arises because the growth of fluctuations is inevitable from the beginning of time.

Consider the history of a fluctuation destined eventually to form a galaxy. It grew in amplitude throughout much of the early expansion of the Universe. The earliest epoch that the cosmologist can usefully discuss is the Planck instant, a mere 10^{-43} second after the singularity. At this time,

Originally published as "Origin of the Galaxies," Joseph Silk, *Nature*, 292, 409–411, 30 July 1981.

our galaxy-sized fluctuation must have had an infinitesimal but non-zero density contrast. These inhomogeneities are best considered as fluctuations in the spatial curvature, or slight wrinkles in the geometry of space-time. The magnitude of such primordial wrinkles must be about 1 part in 10^5, if fluctuations in the energy density are considered; for fluctuations in the matter component alone, the associated wrinkles attain a considerably smaller but well defined value. Were the fluctuation amplitudes much smaller, galaxies would not have formed by now; were they much larger, eventual collapse in the early Universe would have formed separate topologically disjoint universes, bearing little relation to our observed universe.

Although this argument purports to show that the initial conditions may have been unique, it certainly does not provide any explanation. One is left with the following conundrum. The very early Universe was uniform to a high degree. However, to form galaxies, fluctuations must have been present at an exceedingly small but non-zero level. This situation is aesthetically unappealing—granted that the early Universe must have been very nearly uniform, one would obviously prefer to begin with a perfectly homogeneous and isotropic expansion, and see fluctuations develop spontaneously at some later epoch. No one has been able to discover a means of generating fluctuations over a sufficiently large scale to form galaxies in the standard big bang cosmology. To some cosmologists, this failure has provided sufficient reason to reject the standard model. Some have opted for an initially cold big bang in the hope that cold matter may be more susceptible to fragmentation. Others have taken a stronger stance, and rejected the big bang entirely.

A fascinating resolution of this cosmological dilemma was proposed by Ya. B. Zel'dovich[1] of the Space Research Institute, Moscow, and Alexander Vilenkin[2] of Tufts University. Applications of elementary particle theory to the very early Universe imply the possibility of a novel mechanism for the spontaneous generation of the seed fluctuations from which galaxies eventually formed. The origin and nature of elementary particles can be understood in terms of a class of theories known as the grand unified theories (GUTs), which seek to unify the electromagnetic, weak and strong nuclear forces into a single theoretical framework. Above the grand unification energy of about 10^{15} GeV, these fundamental forces of nature all play an equal part, and matter is completely symmetrical in all its properties. The very early Universe provides a natural laboratory for testing GUTs, since particle energies attained then greatly exceed those produced in any manmade particle accelerator. One of the most elegant consequences of GUTs is that they predict (at least in order of magnitude) the observed baryon asymmetry of the Universe. The breakdown of symmetry occurs some 10^{-35} second after the big bang, when particle energies have dropped below the grand unification energy. This produces a slight excess of particles over antiparticles which is crucial to the existence of an environment in which life could evolve, for the fact that we and our surroundings consist exclusively of matter rather than antimatter implies that in the very early Universe, when the cosmic blackbody radiation copiously produced particle-antiparticle pairs, the amount of matter exceeded that of antimatter by about 1 part in 10^9. Our very existence is therefore a consequence of GUTs and the breakdown of symmetries as the Universe expanded and cooled, for a Universe which precisely conserved matter-antimatter symmetry would have undergone considerable annihilation; little residual matter (or antimatter) would remain. Historically, this result was important, since it applied an important idea, originally due to Andrei Sakharov, to a specific epoch in early universe cosmology. A modern alternative interpretation appeals to a later phase transition, associated with the breaking of symmetry between the weak nuclear force and electromagnetism at about 200 GeV, to generate baryon asymmetry.

Zel'dovich and Vilenkin argue that GUTs may have profound implications for the large-scale structure of the Universe. The breakdown of symmetry at 10^{-35} second can be regarded as a phase transition. A useful analogy is an isotropic ferromagnet, which when cooled below a critical temperature (the Curie point) spontaneously develops magnetism. The direction of magnetization is indeterminate and random, and may vary in different regions or domains. The GUTs phase transition may involve the spontaneous generation of a domain structure. The symmetry breakdown is believed to be effected via an intermediary class of particle—the Higgs particle—and its associated fields. These are important in the unification of strong and electroweak interactions, being responsible for the masses of the elementary particles once the symmetrical state is broken. As the Universe cools, the symmetry breakdown occurs because particles and radiation are no longer energetic enough to create the full panoply of states of all types of particle predicted by GUTs. The end state of the matter is characterized by a vacuum state of lowest energy. In the classical physics regime, the vacuum has approximately zero energy, although small fluctuations are present. The fluctuations are variations in the energy of a reservoir of virtual pairs of particles and antiparticles that are released under extreme conditions of temperature or gravitational stress. The vacuum state after the GUTs phase transition has a finite energy density associated with the Higgs field. It is believed that the Universe began with a very large vacuum energy density of almost equal magnitude but opposite sign, forming a false vacuum state, so that one ends up after the phase transition in the true vacuum of zero energy.

In fact, the true vacuum state is not unique. Although the minimum energy state of the Higgs field has a universal value, implying the same energy density everywhere, the direction of the associated field is indeterminate. Multiple true vacuum states are possible, depending on the degree of complexity of the Higgs field. The phase transition can result in the formation of different domains of vacua, corresponding to the multiple vacuum states. The direction but not the magnitude of the Higgs vacuum expectation value differs in each domain. The domain boundaries are characterized by gradients in the Higgs field, and therefore contain energy density; they also consist of residual false vacuum that has not undergone the phase transition. When the Higgs particles have decayed, the Universe contains the remnants of these boundaries in the form of topological structures of false vacuum that may contain a very considerable energy density.

Theory indicates that possible forms for the domain boundaries are topological discontinuities that can be either two-, one-, or zero-dimensional in character. There are two-dimensional walls, one-dimensional strings, or monopoles, which are point-like singularities. An important characteristic of the energy density associated with the monopole's structure is that it effectively behaves as a rest-mass, and assumes a progressively more important gravitational role as the radiation is redshifted away. The walls and strings suffer very little friction from the ambient radiation and matter, and the tendency of these initially highly convoluted structures to straighten out causes them to move at near the speed of light on the horizon scale. Density fluctuations are generated on and below the horizon scale, and larger and larger fluctuations develop as the horizon scale grows.

In fact, expanding walls can be excluded by simple astrophysical considerations. Zel'dovich, Kob'zarev and Okun pointed out[3] that such walls should be extremely massive, and would grossly perturb the cosmic microwave background radiation. However strings are quite another matter. Their one-dimensional structure implies that their influence in the present epoch is relatively modest. Strings may still lead to sizable density fluctuations. The process of generating density fluctuations continues after the phase transition at 10^{-35} second, perhaps until the present epoch if the strings can survive. At formation, the strings are very tangled, and both infinitely long strings and closed loops are present. Dissipation tends to reduce the tension in the strings on scales smaller than the horizon, and the strings gradually straighten out. New strings are likely to form as intersecting strings form closed loops; this provides a principal means of straightening the strings. The expansion of the Universe stretches the strings on scales larger than the horizon, and the characteristic dimension of a string at any epoch is in fact the horizon scale.

The density fluctuations generated by the strings can provide the seeds for galaxy formation. The mass of a galaxy is first contained within the horizon at an epoch of about 10 yr and the mass of a galaxy cluster after about 10^4 yr. Strings will be present on the horizon scale moving at light speed and their gravitational pull will induce density fluctuations. Zel'dovich argues that the density of infinite strings remaining from the GUTs phase transition amounts to about 0.1 per cent of the mean cosmological density. This implies that the expected amplitude of density fluctuations is of the order of 0.1 per cent. Such fluctuations survive the radiation even when their growth is inhibited by the frictional effect of radiative viscosity. After the decoupling epoch, the fluctuations grow in amplitude uninhibited by the radiation, accreting the surrounding matter. Galaxy formation eventually occurs once the fluctuations become of very large amplitude and collapse.

How inevitable is such a scenario? It seems that in general types of GUT, these topological singularities (either walls, strings or monopoles) will form as the symmetry breakdown occurs, depending on the degree of complexity of the Higgs field. The only requirement is that the Higgs field possess a finite correlation length, beyond which its expectation value in different locations is completely uncorrelated. This provides our best hope for understanding galaxy formation in an initially homogeneous and isotropic universe. Galaxies may therefore have originated from a tangle of strings spontaneously produced some 10^{-35} second after the big bang. The strings are best visualized as topological knots of high energy density that may be either of infinite length or closed loops. As the Universe expands, the strings untangle, continuously generating sizable density fluctuations via their gravitational effects on scales less than the instantaneous horizon scale.

Cosmologists are not entirely convinced by this scheme, since it involves rather specific assumptions about elementary particle theory concerning the nature of the Higgs field that have not received widespread acceptance. One difficulty is that not only strings are produced by the GUTs phase transition. Massive monopoles are predicted by the simplest GUTs, and these particles could dominate the present mass density of the Universe, contrary to observation, unless the theory of the very early Universe is carefully adjusted. Perhaps strings will only meet universal acceptance if more direct proof of their existence is forthcoming. This is most likely to emerge from their effect on electromagnetic radiation, in the large-scale angular anisotropy of the cosmic microwave background radiation. Another intriguing possibility is that some of the gravitational lenses, identified with close pairs of quasars having strong spectral similarities, may be due to the gravitational deflection of light by cosmological strings.

---

1. Zel'dovich, Ya. B. Mon. Not. R. Astr. Soc. 192; 663, 1980.

2. Vilenkin, Alexander. Phys. Rev. Lett. 46; 1169, 1981.

3. Zel'dovich, Kob'zarev and Okun. Zh. Eksp. Teor. Fiz. 67; 3, 1974

The Origin of Galaxies

WITH MARTIN J. REES

Perhaps the most startling discovery made in astronomy this century is that the Universe is populated by billions of galaxies and that they are systematically receding from one another, like raisins in an expanding pudding. If galaxies had always moved with their present velocities, they would have been crowded on top of one another about 10 billion years ago. This simple calculation has led to the cosmological hypothesis that the world began with the explosion of a primordial atom containing all the matter in the Universe. A quite different line of speculation argues that the Universe has always looked as it does now, that new matter is continuously being created and that new galaxies are formed to replace those that disappear over the "horizon."

For either hypothesis, it is still necessary to account for the formation of galaxies. Why does matter tend to aggregate in bundles of this particular size? Why do galaxies comprise a limited hierarchy of shapes? Why do spiral galaxies rotate like giant pinwheels? Astrophysicists are trying to answer these and similar questions from first principles. The goal is to explain as many aspects of the Universe as one can without invoking special conditions at the time of origin. In what follows we shall assume a cosmological model in which the Universe starts with a "Big Bang." Some form of continuous creation of matter may not be ruled out with absolute certainty, but there is every indication that the Universe once was far denser and hotter than it is today.

Originally published as "The Origin of Galaxies," Martin Rees and Joseph Silk, *Scientific American*, 26–35, June 1970.

Before the invention of the telescope the unaided human eye could see between 5,000 and 10,000 stars, counting all those visible in different seasons. Even modest telescopes revealed millions of stars and in addition disclosed the existence of many diffuse patches of light, not at all like stars. These extragalactic "nebulas," many of them beautiful spirals, are seen in all directions and in great profusion. As early as the 18th century, Sir William Herschel and Immanuel Kant suggested that these nebulas were actually "island universes," huge aggregations of stars lying far beyond the limits of the Milky Way.

The validity of this hypothesis was not confirmed until 1924, when the American astronomer Edwin P. Hubble succeeded in measuring the distances to a number of spiral nebulas. Several years earlier Henrietta S. Leavitt had shown that Cepheid variables, named for the prototype Delta Cephei, a variable star discovered in 1784, had light curves that could be correlated with their magnitude. The distances of a number of Cepheids were later determined by independent means, so that it became possible to use more distant Cepheids as "standard candles" to establish a distance-magnitude relation. Hubble looked for Cepheid variables in some of the nearer external galaxies and found them. From their period he was able to deduce their absolute luminosity, and from this he was able to estimate their distance. Hubble soon established that the nearest spiral nebulas (or galaxies) were vast systems of stars situated a million or more light-years outside our own galaxy.

Subsequently, Hubble developed a scheme for classifying galaxies according to their morphology, ranging from systems that are amorphous, reddish and elliptical to systems that are highly flattened disks with a complex spiral structure containing many blue stars and lanes of gas and dust. The spiral galaxies themselves vary in appearance. At one extreme are those with large, bright nuclei and inconspicuous, tightly coiled spiral arms. At the other extreme are galaxies in which the nuclei are less dominant and the spiral arms are loosely wound and prominent. The elliptical galaxies also form a sequence, ranging from almost spherical systems to flattened ellipsoids. In addition there are highly irregular systems showing very little structure of any kind.

In all these sequences there is a parallel progression in certain characteristics of the galaxies. In general, spirals are rich in gas and dust, contain many blue supergiant stars, are highly flattened and rotate appreciably. Ellipticals, by contrast, seem to possess little gas or dust, usually contain late-type dwarf stars and exhibit scant rotation.

The masses of galaxies are found by several methods. Galaxies are often gravitationally bound together in pairs. If the distance between them and their relative velocities are known, Kepler's law can be used to find their total mass. Another method, used mostly for spirals that are viewed edge-on or obliquely, is to determine the velocity of rotation by measuring the Doppler shift of spectral lines emitted by ionized gas in various parts of the disk. (The spectral lines of approaching gas will be shifted toward the blue end of the spectrum, those of retreating gas toward the red end.) One can plot a rotation curve showing how the velocity of rotation varies with the distance from the center of the galaxy. The mass can then be estimated from the requirement that the centrifugal and gravitational (centripetal) forces must be in balance. It turns out that the masses of galaxies are typically about 10^{11} (100 billion) times the mass of the sun. The range, however, is fairly broad: from about 10^{8} solar masses for some nearby dwarf galaxies to 10^{12} solar masses for giant ellipticals in more remote regions of the Universe. The diameter of the larger spirals, such as our own galaxy, is about 100,000 light-years.

Galaxies also differ widely in their ratios of mass to luminosity. Taking the mass-to-luminosity ratio of the sun as unity, one finds that within the optical confines of large spirals, such as our own galaxy, the ratio varies from one up to 5. In other words, some spirals emit only a fifth as much light per unit of mass as the sun does. Ellipticals commonly emit even less: only about a twentieth as much light per unit of mass. (Thus their mass-luminosity ratio is 20.)

The distribution of galaxies in the sky is quite inhomogeneous. There are many small groups of galaxies, and here and there some rich clusters containing up to 1,000 members or more. Such systems vary from one million light-years across to 10 million. Our own galaxy is a member of the "local group," an association of about 20 galaxies, only one of which, the Andromeda galaxy, has a mass comparable to that of ours. The local group is about three million light-years in diameter. The Andromeda galaxy is some two million light-years away; the nearest large cluster of galaxies is in Virgo, about 50 million light-years distant.

Even such clusters do not seem to be randomly distributed in space. Some astronomers have argued that there is evidence that clusters are grouped into superclusters of perhaps 100 members, spread over 100 million light-years. The Universe appears to be uniform on scales larger than this.

Establishing the distance of galaxies was only part of Hubble's achievement. Working with the 100-inch telescope on Mount Wilson, he showed from red-shift measurements that the galaxies are in recession.

Hubble found, moreover, that the red shift of a galaxy is directly proportional to its distance, as judged by its apparent luminosity. In the 1970's, the most distant galaxies known were in a faint cluster in the constellation Boötes; Rudolph Minkowski discovered that the wavelength of light coming from this cluster is stretched by 45 percent. The redshift is 0.45, and the corresponding velocity of recession is nearly half the speed of light. The most distant galaxy discovered by 1993 is at a redshift of 3.8; the wavelength of light is lengthened by 380 percent or measured by the astronomer. Light originating from some of the brilliant starlike objects known as quasars is red-shifted by nearly 500 percent.

The light from Minkowski's cluster of galaxies set out toward us about five billion years ago, and so we can be sure that some galaxies are even older than that. On the other hand, as we have mentioned, all the galaxies must have been tightly packed together no more than 10 billion years ago, based on their present recession velocity. Estimates of the ages of stars suggest that our galaxy, and others like it, are unlikely to be much less than 10 billion years old. Hence we are presented with a remarkable coincidence: most galaxies appear to be about as old as the Universe. This implies that galaxies must have formed when conditions in the Universe were much different from those now prevailing.

It seems clear, then, that the formation of galaxies cannot be treated apart from cosmological considerations. The dynamics and structure of the Universe in the large are beyond the scope of Newtonian physics; it is necessary to use Einstein's general theory of relativity. Because of the complexity of the theory, it is practicable to solve the equations only for cases having special symmetry. Until quite recently the only solutions for an expanding Universe were those found in 1922 by the Russian mathematician Alexander A. Friedmann. In his idealized models matter is treated as a strictly uniform and homogeneous medium. The Universe expands from a singular state of infinite density, with the rate of expansion decelerating as a consequence of the mutual gravitational attraction of its different parts. The Universe may have enough energy to keep expanding indefinitely, or the expansion may eventually cease and be followed by a general collapse back to a compressed state. Observations of the actual rate of expansion of the Universe at different epochs, as determined by the red shift-luminosity relation of the most distant galaxies, fail to tell us unambiguously whether the expansion will finally stop and be reversed or whether it will continue indefinitely.

The clumping of matter into stars, galaxies and clusters of galaxies in the real Universe might seem to make Friedmann's models, based on perfect homogeneity, empty exercises. In actuality, the "graininess" we observe in the Universe is on such a small scale that Friedmann's solutions remain valid. The reason is that the gravitational influence of local irregularities is swamped by that of more distant matter.

Perhaps the most convincing evidence in support of Friedmann's simple description of the Universe was supplied in 1965 by the discovery that space is pervaded by a background radiation that peaks at the microwave wavelength of about two millimeters, corresponding to the radiation emitted by a black body at an absolute temperature of three degrees (three degrees Kelvin). This radiation could be the remnant "whisper" from the Big Bang of creation. The remarkable isotropy, or nondirectionality, of this radiation is impressive evidence for the isotropy of the Universe.

The radiation was discovered independently and almost simultaneously at the Bell Telephone Laboratories and at Princeton University. The radiation has the spectrum characteristic of radiation that has attained thermal equilibrium with its surroundings as a result of repeated absorption and reemission, and it is generally interpreted as being a relic of a time when the entire Universe was hot, dense and opaque. The radiation would have cooled and shifted toward longer wavelengths in the course of the universal expansion but would have retained a thermal spectrum. It thus constitutes remarkably direct evidence for the hot-Big-Bang model of the Universe first examined in detail by George Gamow in 1940.

Assuming the general validity of the Friedmann model for the early stages of the Universe, it seems clear that the material destined to condense into galaxies cannot always have been in discrete lumps but may have existed merely as slight enhancements above the mean density. There will be a tendency for the larger irregularities to be amplified simply because, on sufficiently large scales, gravitational forces predominate over pressure forces that tend to oppose collapse. This phenomenon, known as gravitational instability, was recognized by Newton, who, in a letter to Richard Bentley, the Master of Trinity College, wrote:

"It seems to me, that if the matter of our sun and planets, and all the matter of the Universe, were evenly scattered through all the heavens, and every particle had an innate gravity towards all the rest, and the whole space throughout which this matter was scattered, was finite, the matter on the outside of this space would by its gravity tend towards all the matter on the inside, and by consequence fall down into the middle of the whole space, and there compose one great spherical mass. But if the matter were evenly disposed throughout an infinite space, it could never convene into one mass; but some of it would convene into one mass and some into another, so as to make an infinite number of great masses, scattered great distances from one to another throughout all that infinite space. And thus might the sun and fixed stars be formed, supposing the matter were of a lucid nature."

Newton envisaged a static universe, but the same qualitative picture occurs in an expanding Friedmann universe, as was shown by the Russian physicist Eugene Lifshitz in 1946.

Because of the atomic nature of matter, the early Universe could never have been completely smooth. It would obviously be gratifying if the inevitable random irregularities in the initial distribution of atoms sufficed ultimately to produce the bound systems of stars we see throughout the Universe today. Unfortunately this type of statistical fluctuation fails by many orders of magnitude to account for the observed degree of structure in the Universe. Moreover, it remained a puzzle why agglomerations of a certain mass, notably galaxies, should be so plentiful. It appeared necessary to postulate initial fluctuations in a seemingly ad hoc manner, and nothing had really been explained; "things are as they are because they were as they were."

In the late 1960's, it was determined that the background radiation acts as a gigantic homogenizer on certain preferred scales. To understand just how this works we must look more closely at Gamow's model of the universe. In the early stages, when the Universe consisted of a primordial fireball, no structures such as galaxies or stars could have existed in anything like their present form. All space would have been filled with radiation (photons) and hot gas, consisting of the nuclei of hydrogen and helium and their accompanying electrons. The photons would be repeatedly scattered from the electrons. For at least the first 100,000 years of its history (beginning roughly 10 seconds after its emergence from the initial singularity) the Universe can be pictured as a composite gas in which some of the "atoms" are particles and the rest are photons. For the Universe as a whole there are now at least 100 million times more photons than particles. From thermodynamic considerations, one can conclude that photons must also have greatly outnumbered particles in the fireball. For a gas in equilibrium, each species of particle contributes to the total pressure in proportion to its number. This still holds (very nearly) for photons, so that the radiation would make an overwhelmingly dominant contribution to the pressure. (During the first seconds, when the temperature exceeds a few billion degrees, the situation is less simple because pairs of photons can interact to form an electron and a positron.)

As the expansion proceeds and the density decreases, the photons lose energy, the temperature drops and the particles move less rapidly. A key stage is reached after about 3×10^5 years, when the fireball has cooled to 3,000 degrees. The electrons are then moving so slowly that virtually all are captured by nuclei and retained in bound orbits. In this condition they can no longer scatter photons and the Universe becomes transparent. Inasmuch as the background temperature today is only about three degrees absolute, one can conclude that the Universe has expanded by a red-shift factor of 1,000 since the scattering stopped. (Wavelength is inversely proportional to temperature.)

The microwave background photons have probably propagated freely since the Universe became transparent and therefore they should carry information about a "surface of last scattering" at a red shift of more than 1,000. Compare this with the red-shift factor of about one for the most distant galaxies known! Because these photons have been traveling unimpeded since long before galaxies existed, they should provide us with remarkably direct evidence of physical conditions in the early Universe.

Let us return now to the epoch of the primordial fireball and ask: How were inhomogeneities in the fireball affected by the presence of the intense radiation field? Radiation would inhibit the process of gravitational collapse. Under radiation pressure, nonuniformities in the fireball would take the form of oscillations, pressure waves or turbulence. These disturbances, in turn, will be dissipated by viscosity and the development of shock waves. Some wavelengths will be attenuated more severely than others, so that inhomogeneities of favored size will be preserved whereas those less favored will tend to be destroyed. The aim of recent work has been to determine what scales of perturbation are most likely to survive the various damping processes until the scattering of photons comes to an end. Any perturbation whose survival and growth is specially favored should eventually dominate, almost irrespective of how nonuniformities were initially distributed in the primordial fireball. An encouraging result that has already emerged from these studies is that between 10^{12} and 10^{14} solar masses, roughly the mass of a large galaxy or a small cluster of galaxies, is one such preferred scale.

After the electrons in the initial plasma have been bound into atoms, radiation no longer affects the distribution of mass. At this point, the surviving perturbations are free to amplify gravitationally. (It should be noted, however, that on small scales—less than 10^6 solar masses—the kinetic energy of atoms exerts a pressure of its own that inhibits gravitational collapse.) The first generation of bound systems will therefore condense from whatever scale of fluctuations had the largest amplitude at the time of decoupling; that is, when the fireball ceased to be a plasma of electrons and other particles.

At what stage did protogalaxies stop expanding and separate out from the rest of the Universe? We might guess that this happened when the mean density was comparable to the present density in the outlying parts of galaxies. In 1962, Olin J. Eggen, Donald Lynden-Bell and Allan R. Sandage of the Hale Observatories investigated the likely early history of our own galaxy by studying very old stars in the galactic halo. These stars probably formed while the galaxy was collapsing to its present disklike shape (and before the birth of the stars in the Milky Way), and their orbits indicate that our galaxy attained a maximum radius of about 100,000 light-years. One can then tentatively estimate that galaxies such as our own formed when the Universe was 1,000 times denser than it is now, about half a billion years after the expansion began.

Extrapolating backward in time, we find that the protogalaxies would have taken the form of nonuniformities roughly 1 percent denser than the average density of the Universe at the decoupling epoch. It is an attractive possibility that these are the dominant surviving irregularities, all smaller scales having been smoothed out during the fireball phase. There are, however, some types of fluctuation that are not eradicated in the fireball, so that smaller gas clouds may have formed first and later collided and agglomerated into galaxies. Robert H. Dicke and P. J. E. Peebles of Princeton have suggested that globular clusters—compact groups of about 10^5 or 10^6 stars that orbit around galaxies—may represent that small fraction of clouds which managed to avoid collisions, fragmented into stars and survived. Clusters of galaxies would have evolved from initial irregularities of smaller amplitude but larger scale than those destined to form single galaxies.

The main contribution of cosmologists to date toward explaining galaxy formation has been to calculate what scales of perturbation are most likely to survive or amplify in the fireball, thereby reducing the need to build these preferred scales into the initial conditions. This removes one element of arbitrariness in the initial conditions prescribed for the Universe. There still remains, however, the task of explaining both the origin of the nonuniformity of the Universe on all scales except the very largest, and the apparent uniformity encountered on the largest scales.

In fact, the Friedmann models may not provide an adequate description of the fireball when large inhomogeneities are present. It would be conceptually attractive if there were processes that could transform an initially chaotic universe into one that displayed the large-scale uniformity of a Friedmann model. An encouraging step toward this goal has been taken by Charles W. Misner of the University of Maryland, who has considered a "mixmaster" universe, which expands anisotropically in such a way that all parts of the Universe are causally related very early in its history. At the outset, matter would be so densely packed that even neutrinos would interact with other particles at a significant rate. Acting like a blender, the neutrinos would destroy the original anisotropy of the fireball by the time it had cooled to about 20 billion degrees. Thereafter the expansion would mimic a homogeneous Friedmann model.

Several types of observation may help to test this general picture of galaxy formation. The fluctuations that develop into galaxies and clusters would give rise to random motions on the surface of last scattering. As a result, the microwave background photons would not all have been redshifted by exactly the same amount; in some directions they might have been scattered off material with a random velocity toward us, whereas in other directions the last-scattering surface may have been receding from us. As a consequence the microwave temperature would be slightly nonuniform over the sky. Radio wave, millimeter wave and submillimeter wave observers can detect temperature fluctuations as small as one-thousandth of a percent on angular scales of a few minutes of arc, an inverse of a hundredfold over the detection limit 20 years ago at radio wavelengths. So far, they have found no positive effect, although fluctuations have been recently reported as degree scales. This technique, however, has the potentiality of detecting embryonic galaxies or clusters of galaxies when they were merely small enhancements above the mean gas density.

There are reasons to expect galaxies that have just condensed to be brighter than typical galaxies at the present epoch. The energy released by the collapse of the protogalaxy would probably have been radiated away by hot gas before most of the stars formed. Moreover, the first generation of stars would tend to be heavier and more luminous in relation to their mass than the stellar populations in present-day galaxies. Although most of this energy would be radiated in the ultraviolet, it would be received in the near infrared, owing to the red shift. Robert Bruce Partridge and Peebles at Princeton have suggested that it might be feasible to detect such young galaxies even though these may now have red shifts of about 10.

We are plainly still far from understanding even the broad outlines of the processes whereby the observed aggregations of matter in the Universe came into being. We are even further from understanding the detailed morphology of the bewildering variety of different types of galaxies. For example, we have not yet discussed the possible origins of the angular momentum or magnetic fields of galaxies. Peebles has argued that the rotation of galaxies may be induced by tidal interactions soon after formation. Other authors, notably Leonid Ozernoi of the P. N. Lebedev Physical Institute in Moscow, have considered galactic rotation to be of primordial origin. One remarkable feature of the primordial fireball is that it can store rotation in the form of "photon whirls"; subsequently this stored rotation could be transferred to matter whirls. Such whirls, however, result in gravitational field fluctuations that are very large at early times, and would wreck the predictions of light element synthesis, as well as produce excessively large fluctuations in the cosmic microwave background in small angular scales.

Galactic magnetic fields may be produced after the formation of the galaxy by a mechanism of the dynamo type. Alternatively, magnetic fields may very well be primordial origin. Edward R. Harrison of the University of Massachusetts has pointed out that the shear between the photon gas and the matter gas in the fireball could have generated a small magnetic field; if primordial photon whirls are assumed to be present, this mechanism leads to the production of a "seed" magnetic field many orders of magnitude below the value of the magnetic field observed in the spiral arms of our own galaxy. Harrison argues that rapid rotation of the protogalaxy may have subsequently produced sufficient winding of the primordial magnetic field to enhance it by the dynamo mechanism to the field currently observed. Primordial fields alone would be insufficient to account for the observed galactic fields. The amount of rotation and the strength of the magnetic field in the protogalaxy probably help to determine whether it will evolve into an elliptical galaxy or into a spiral.

Galaxies are observed to possess random velocities with respect to the cosmic expansion. It is a curious coincidence that the rotational velocity of galaxies is of just the same order of magnitude—hundreds of kilometers per second—as these random motions. Perhaps this is simply a consequence of primordial origin for rotation: tidal interactions drive shearing motions as well as induce vorticity, as long as the overall angular momentum is conserved.

Present data on the sizes of clusters of galaxies, and on possible "superclusters," are too sparse to enable us to assess the validity of theories that predict the mass spectrum of condensations. Moreover, our knowledge of the masses of galaxies is bedeviled by selection effects. Large and bright galaxies can be seen out to great distances, but small and intrinsically faint ones would only be noticed if they were comparatively close to us. Such objects may therefore occur much more frequently than is believed. A more drastic possibility is that most of the material in the Universe may be in some nonluminous form. Evidence for the existence of such material comes from studies of the stability of clusters of galaxies.

This basic problem was first discussed in 1933 by Fritz Zwicky of the California Institute of Technology. For example, if one estimates the mass required to make the Virgo cluster a gravitationally bound system, one finds that the total observed mass in the member galaxies falls short by a factor of 50 or more. One possible way around this paradox is to assume that the Virgo system may be exploding, as the Soviet astrophysicist V. A. Ambartsumian has suggested. Perhaps even more puzzling is the apparent deficiency in mass of the Coma cluster. This system is so spherically symmetric and centrally condensed that astronomers believe it must be a stable system. Yet the observed mass, predominantly in elliptical galaxies, falls short of the mass required for stability by perhaps a factor of five, even if one assumes that the mass-luminosity ratio for ellipticals is around 50.

Similar results have been found for other clusters. These systems clearly are long-lived and centrally concentrated; they may be stable. Astronomers have attempted to explain this problem by arguing that nonluminous matter is present in sufficient quantity to stabilize these systems. This material probably cannot all be in gaseous form; neutral hydrogen or ionized hydrogen, whether uniformly distributed or in clouds, ought to be observable either by radio or by X-ray observations.

Alternatively, the "missing mass" may be in the form of low mass stars are stellar remnents, possibly in intergalactic space or in "dead," or burned-out, galaxies. An even more intriguing possibility is that concealed within the clusters are many objects that have undergone catastrophic gravitational collapse, as predicted by the general theory of relativity. The gravitational field around such objects would be so strong that no radiation could escape from them; only their gravitational influence could be detected by a distant observer.

Other arguments that indicate the apparent youthfulness of some galaxies stem from observations of clusters of galaxies. Distant clusters are teeming with very blue, star-forming galaxies, unlike their nearby counterparts. One seems forced to the conclusion that here are newly formed galaxies, born within the past billion years. Zwicky has discovered an entire class of star-forming compact galaxies whose surface brightness resembles that found only in the nuclei of ordinary galaxies. Even more baffling is the discovery that some quasars emit as much radiation as 1,000 galaxies, the energy apparently coming from a colossal explosive event in a region less than 1 percent the size of the solar system. Seyfert galaxies display the same energetic phenomenon on a somewhat reduced scale.

Ambartsumian has long maintained that galactic nuclei are sources of matter and that indeed the galaxies themselves emerge out of dense primordial nuclei. In recent years Halton C. Arp of the Hale Observatories and Erik B. Holmberg of the University of Uppsala have found evidence that small galaxies may even have been ejected from larger galaxies. These phenomena certainly suggest that violent events, involving perhaps the birth of galaxies, are continually taking place in the nuclei of existing galaxies. One is reminded of Sir James Jeans's prescient conjecture, written in 1929, that "the centers of the nebulae are of the nature of 'singular points,' at which matter is poured into our Universe from some other, and entirely extraneous, spatial dimension, so that, to a denizen of our Universe, they appear as points at which matter is being continually created."

But these are very much minority viewpoints. Quasars are identified as the anomalously bright centers of otherwise normal galaxies. Seyfert galaxies are a less energetic phenomenon, but similarly involve the bright nucleus of a galaxy. There seems to be a continuum of activity that stretches from quasars to low luminosity active galactic nuclei and even to the nuclei of galaxies like the Milky Way, where there is evidence of past energetic outbursts. It is commonly believed that a central massive black hole, weighing between 10^6 and 10^9 solar masses, formed during the early phase of galaxy formation. Subsequent accetion, of gas shed by evolving stars, and the debris resulting from stellar collisions by stars that ventured too near the Schwarzschild radius, fuels the black hole and is responsible for the luminous activity. In truth, there is a monster lurking in the middle.

Further progress in this field must await fuller information on the distribution, masses and velocities of galaxies. Moreover, satellite observations in infrared, ultraviolet and X-ray wavelengths may soon reveal completely new and unsuspected types of objects, and should in any case give us confidence that we have a fairly complete inventory of the contents of the Universe. We shall then be better able to relate theoretical abstractions to the Universe in which we dwell.

NEAR AND FAR

Moving as One

No matter how hard cosmologists try, they have been unable to come up with a simple scenario that simultaneously explains both galaxy formation and the large-scale frothy structure of the galaxy distribution. One class of models, characterized by cold dark matter in the form of weakly interacting elementary particles, does well at explaining small-scale structures, but stumbles at the hurdle of the clustering of galaxy clusters. Another class, characterized by hot dark matter in the form of massive neutrinos or dark baryons such as burnt-out stars or black holes, shows promise of matching the large-scale structure, but at the cost of failing to form galaxies early enough in the Universe. The one point of consensus is the existence of pervasive dark matter, although the lack of any evidence as to its nature has opened the floodgates of speculation. More exotic avenues are being explored, the most promising of which involves cosmic strings, relics from the very early Universe which can act as seeds for the growth of structure on both small and large scales.

It is premature to tell whether strings will be the cosmic panacea. But no sooner have cosmologists recovered from the momentous discovery of the bubble-like structure of the galaxy distribution on scales of tens of megaparsecs (Mpc) than a remarkable new result is being announced. That our Local Group of galaxies is whizzing through space at a speed of about 600 km s^{-1} towards a direction not far from the southern constellation of Centaurus is well known. This motion shows up as an unambiguous dipole anistropy in the cosmic microwave background (CMB) radiation, seen as a slight heating of the radiation in the direction in which we are moving and as a slight cooling in the reverse direction. Recently it has been reported that our Local Group is not a lone wanderer in intergalactic

Originally published as "What Makes Nearby Galactic Clusters All Move as One?" Joseph Silk, *Nature*, 322, 207, 17 July 1986.

space, but that galaxies throughout a vast region of about 100 Mpc in extent are companions in its headlong rush. Evidence for this alarmingly coherent large-scale flow has been independently found by two groups.

David Burstein (Arizona State University) and co-workers from six other institutions spanning the globe from Pasadena to Herstmonceux (R. Davies, A. Dressler, S. Faber, D. Lynden-Bell, R. Terlevich and G. Wegner) have presented data on some 400 elliptical galaxies evenly distributed on the sky. The measured parameters (central velocity dispersions, effective radii and total magnitudes) defined a sufficiently good correlation that enable distances to be inferred once distance-independent parameters (such as velocity dispersion and total magnitude) have been measured. This survey shows that there is a bulk motion of about 700 km s^{-1} of essentially all galaxies within 60 h^{-1} Mpc of the Local Group ($h = 1$ if the Hubble constant is 100 km s^{-1} Mpc^{-1}, or $h = 1/2$ if the Hubble constant is half of this value). The direction of this motion is towards galactic longitude $l = 299°$ and latitude $b = +1°$ and within 20° from the apex of the dipole motion which yields the Local Group motion relative to the CMB.

Another group, working at Imperial College London, obtained infrared photometry for an all-sky sample of the class of galaxies termed Sc at a mean distance of 50 h^{-1} Mpc. Again, by adopting the known correlations between their derived distance-dependent parameter (infrared luminosity) and distance-independent parameters (infrared colour or central velocity dispersion) either absolute distances or Hubble velocities can be inferred once the correlations are calibrated. Comparison of Hubble velocity with the observed recession velocity for about 45 galaxies led the group to infer a bulk streaming velocity of 970 (±300) km s^{-1} in a direction towards $l = 305°$, $b = 47°$ that, within their large quoted uncertainty, is consistent with Burstein et al.'s result.

To many astronomers, this is a case of déjà vu. The large-scale motion amounts to a confirmation of the Rubin–Ford effect, a motion of the Local Group relative to a shell of Sc galaxies at ~50 h^{-1} Mpc that, following its initial discovery in 1976 by Vera Rubin and Kent Ford, has generally been disbelieved until now. After all, the inferred direction of motion was nearly orthogonal to the apex of the Local Group motion relative to the CMB. This neglect may have been unjustified. Indeed, the first indication of a large-scale bulk motion came with the confirmation by several groups that the Local Group is falling towards the Virgo cluster at ~ 250 km s^{-1} and in a direction some 45° away from the apex of its CMB motion. This means that, in the reference frame of the CMB, the entire Virgo Supercluster, including the Local Group at a distance of some 10 h^{-1} Mpc from Virgo, is moving at ~400 km s^{-1} in the direction of Hydra–Centaurus.

The new data on ellipticals tell us that clusters in Hydra–Centaurus in the south, and in Perseus and Pisces in the north, separated by a distance of ~60 h^{-1} Mpc, are all sharing a common motion with Virgo, whose amplitude may be as large as 700 km s^{-1}. We are all moving towards a region behind the Hydra–Centaurus clusters. According to Burstein et al., there is an additional component of random motion of ~300 km s^{-1} for individual clusters superimposed on the bulk flow. Finally, of course, the CMB must represent the ultimate local rest frame. It is reassuring that A. Yahil, D. Walker and M. Rowan-Robinson have confirmed in a study of IRAS galaxies that at a depth of ~200 h^{-1} Mpc, the galaxies are indeed at rest with respect to the CMB.

Reconciling these new data with existing models of large-scale structure will be a difficult task, as no model predicts such large velocities. If the interpretation of the data holds up, then something must be pulling us; exactly what is not clear. Indeed, it has been argued by N. Vittorio, R. Juszkiewicz and M. Davis that confirmation of the large-scale bulk motion will invalidate all hot or cold dark matter models that rely on inflation-generated random phase perturbations of a Friedmann cosmology. Less attractive models may work: these include low-density Friedmann cosmologies containing either hot or cold dark matter, together with primordial seeds that have triggered early galaxy formation. The seeds might be cosmic strings, as proposed by T. Kibble, A. Vilenkin and Ya B. Zel'dovich, or rare objects that have injected sufficient energy via supernova explosions to explosively amplify the perturbed mass scale up to galactic dimensions, as discussed by J. Ostriker and L. Cowie.

The one source of solace is that the large-scale bulk motion does seem to be a likely consequence of the existence of the Hubble bubbles. These apparent voids, on scales up to 50 h^{-1} Mpc, could have been generated by the large-scale flows which would evidently be present in a hot dark matter-dominated universe. Of course, these flows must have maintained considerable coherence to preserve the well-defined bubble surfaces on which the galaxies are found. If this were the case, the bubble interiors should be genuinely devoid of matter.

There is an alternative, needless to say, offered by advocates of cold dark matter—namely, that the voids are illusory, simply reflecting the large-scale inhomogeneity of the luminous matter, concentrated into great clusters and into ridges of galaxies, from the relatively smooth dark matter distribution. Such a situation could arise if only the highest peaks in the primordial fluctuation spectrum managed to form galaxies. But however this biasing arose, one consequence is inevitable: the large-scale flows must turn out to be an artifact of observational error. Astronomers are now rushing to verify the reality of the large-scale flows: their confirmation promises to mark a turning point in cosmology.

Five years after these words were written, the data have greatly improved, with two new independent techniques for measuring bulk flows. One approach uses the Tully-Fisher correlation between the luminosity and rotational velocity of a spiral galaxy. If this is a universal correlation, as measurements suggest, one can measure luminosity, which is distance-dependent, and rotation velocity, which is independent of distance, for a distant galaxy and infer the true distance, and hence peculiar velocity.

A second approach reconstructs the three-dimensional velocity field. The flows persist: towards Centaurus in the south, and now in the new data, from Perseus-Pisces, almost 180 degrees away to the north. The typical velocity is confirmed to be about $400 kms^{-1}$, out to $60h^{-1}$ Mpc. Doubts persist however, since the velocity measurements are indirect, and rest on assuming that properties of galaxies, based on their stellar content, that are used to derive distances are universal and do not vary with location. Nature would perhaps have to be perverse for this assumption to be invalid, but galaxies are fragile systems susceptible to environmental effects. Improved data on the stellar content of galaxies will eventually provide reassurance that we are really measuring peculiar velocities.

Deciphering the Cosmic Code

Take a mixture of gas and dust, cook it with the aid of a large computer, and a galaxy may emerge. That, at least, is the dream of astronomers who study the most remote galaxies. For we see these galaxies as they were many billions of years ago. A youthful galaxy should differ in appearance from a mature galaxy, and an understanding of galactic evolution is crucial to deciphering the observational evidence. In practice, a much more pragmatic approach is adopted. One begins by synthesizing the spectrum of a nearby galaxy, and systematically develops an increasingly sophisticated model which can cope with very distant and possibly younger galaxies. A workshop[1] was largely devoted to modelling galactic spectra.

Stars provide a vital ingredient. C. de Loore (University of Brussels) described how the evolution of massive stars is affected by mass-loss via stellar winds and by convective mixing after their hydrogen fuel is exhausted. Massive stars are indeed observed to be shedding substantial amounts of material at velocities up to several thousands of kilometres per second. Since the rate of evolution of a star depends in a critical way on the gravitational field and central temperature, the implied change in mass can have a large effect on the luminosity and colour of the star as it ages after hydrogen-burning. While the actual mass-loss rates are quite well known, they differ considerably from star to star, even at similar luminosities. The main constraint comes from the Hertzsprung–Russell diagram which interprets the relationship between stellar luminosity and effective temperature in terms of stellar aging. Unfortunately, the spread in mass-loss rates for stars that are otherwise indistinguishable in the diagram allows considerable freedom in choice of mass-loss and convection

Originally published as "Deciphering the Cosmic Code," Joseph Silk, *Nature*, 304, 304–305, July 1983.

parameters, and consequently in stellar enrichment, for example.

But such massive stars have died after some tens of millions of years, their evolution culminating in a fiery supernova explosion. It is stars of lower mass that dominate the light from nearby galaxies, and their properties were reviewed by A. Renzini (University of Bologna) and I. Iben (University of Illinois). As stars exhaust the hydrogen fuel supply in their cores and begin to burn core helium, their outer layers swell up and they become luminous red giants. Eventually mass outflows deplete the outer envelope as the nuclear fuel supply in the stellar core is exhausted and the hot contracting core is exposed as a white dwarf. These processes take ten billion years or longer for stars of solar mass, and could contribute importantly to the blue and ultraviolet light from galaxies.

The low- and intermediate-mass stars account for some of the enrichment of heavy elements. Such old stars can develop into supernovae if a white dwarf is destabilized by accretion of matter from a close binary companion, and prolific production of iron-peak elements occurs in the ensuing collapse and explosion. The massive stars are important producers of oxygen, and observations of oxygen abundance were reviewed by F. Villefond (Observatoire de Paris) and B. Pagel (Royal Greenwich Observatory). Surveys of HII regions, the sites of recent massive-star formation, reveal that in the outer regions of our Galaxy and of other galaxies, systematically lower abundances of oxygen relative to hydrogen are found with increasing galactocentric distance. Simultaneously, the mean effective temperature of the stars increases. This is due either to an increase in the number of massive stars or to an increase in the mass of the most massive stars present. In principle, both effects could be acting, but Villefond argued that the latter effect was more important. He also found that the trend continued in extragalactic HII regions, which include the most metal-poor massive-star-forming system hitherto discovered. Pagel demonstrated that the decrease in oxygen abundance could be correlated with the mass surface density of the galactic disc. The oxygen gradient may be generated either by the incorporation of infall into the galactic disc by primitive metal-poor gas, or by a variable nucleosynthetic yield. In our Galaxy, there is little evidence for infall, although this may not be typical for other systems. But it is possible that the stellar yield of processed matter per unit of mass consumed in star formation and remaining in long-lived stars or compact remnants decreases with galactocentric distance or decreasing surface density; this could explain the oxygen gradient.

The dust component has been measured directly in our Galaxy by extinction studies of bright stars. Ultraviolet observations were compared with similar data for the Magellanic Clouds by K. Nandy (Royal Observatory, Edinburgh), who emphasized the weakness of the 2,200 Å graphite feature in these irregular galaxies. Models indicate that in the Large Magellanic Cloud, graphite is depleted relative to the silicate component, known from the 10 μm feature to be a major grain constituent in our Galaxy, by a factor of 3. In the Small Magellanic Cloud, this depletion is by a factor of 8. Translation of the graphite depletion into information about the carbon abundance is difficult, because, as emphasized by B. Savage (Wisconsin), formation and destruction mechanisms for interstellar grains are poorly understood. This is still a potentially important result, because the major source of carbon is from low-mass stars, whereas massive stars produce oxygen. Variations in the proportion of high- to low-mass stars are among the most important ingredients to incorporate into models of stellar population synthesis (G. Bruzual, Centro de Invetigaciones de Astronomia, Venezuela).

Indeed, Bruzual managed to provide a remarkable synthesis of the spectrum of the light from the centre of the Andromeda galaxy, Messier 31, a region of exclusively old stars. His model incorporates star formation and star death, and can be extrapolated back in time some ten billion years. One or two galaxies, selected by their radio properties, have now been discovered at a redshift larger than one, corresponding to a look-back time of some 10 billion years. Bruzual found that the colours of these remote objects were bluer than those of nearby ellipticals, even when suitable allowance was made for the reddening due to the expansion of the Universe. He interprets this as a sign of spectral evolution. The amount of evolution is modest, but nevertheless signifies star formation in these ellipticals over the first ten billion years of their history. This long period of star formation came as a surprise, since most cosmologists had expected star formation to last no longer than a free-fall time, some hundreds of million years, in an elliptical galaxy.

Some notes of caution, however, were sounded. Radio galaxies are often very luminous galaxies, and may have evolved quite differently from more normal galaxies. S. Lilly (ROE) noted that the infrared colours of the most distant radio galaxies, obtained at the UKIRT telescope, revealed little or no evidence for evolution. This result has the potentially exciting implication that radio galaxies may provide standard candles for use in cosmology, and infrared measurements may lead to a new determination of the cosmological deceleration parameter. Evolutionary effects wreak such havoc with the interpretation of optical data on the redshifts and magnitudes of remote galaxies that a solution does not seem to be forthcoming soon unless a radical new approach is opened up (R. Kron,

Chicago). The IR spectral region could provide such an approach.

Another source of uncertainty in interpreting the optical observations of distant galaxies is that the ultraviolet spectra of nearby ellipticals display considerable variation. Most are flat, indicating few very hot stars; but one or two, including the giant radio galaxy Messier 87, show an increase towards the far ultraviolet (R. Ellis, Durham). Whether this blue light is due to a few newly formed massive stars, as proposed by P. Gondhalekhar (Rutherford), or to an old evolved population of low-mass stars, is uncertain. This uncertainty has considerable implications for the interpretation of the colours of very distant galaxies, where the ultraviolet light is redshifted into visible wavelengths.

An alternative approach to seeking remote luminous galaxies is to perform deep counts of all galaxy images on a photographic plate. Ellis showed that counts obtained by several groups, including those at Durham with the AAT, were now probing cosmological distances. At 22nd magnitude in the blue, one expects to find many galaxies at a redshift of 0.5 or greater. Comparison of the available data suggests that evolution in blue luminosity may be required. The amount of evolution is rather modest, and in fact is very similar to that used by Bruzual in interpreting the colours of the high-redshift radio galaxies. This would again imply that many galaxies were bluer and more luminous in the past.

Where does all this leave us? The evidence for evolution in distant galaxies is growing. Unfortunately, it is impossible to arrive at a unique interpretation of the observations. There may be many possible evolutionary tracks that yield the mature galaxies we now see, and extrapolation into the past becomes exceedingly dangerous while we are still uncertain of what is happening in nearby galaxies and, indeed, in our own Galaxy. We require an improved understanding of the evolution of the youngest stars and of the oldest stars. We must discover how these processes depend on metallicity. We need to understand how star formation is initiated and to measure the initial distribution of stellar masses. And we cannot neglect the effects of galactic environment. Some recent UKIRT observations (N. Robertson, Imperial College, London) show that tidal interactions between galaxies may stimulate star formation. Other environmental effects that may plague distant galaxy observations include those of galactic cannibalism, whereby a luminous galaxy swallows its lesser neighbours, and galaxy mergers.

The goal of using distant galaxies to determine whether the Universe is open or closed seems as remote as ever, but our understanding of galactic evolution is likely to be improved considerably. The challenge posed by theories of galaxy formation is clear. Accumulating evidence from the large-scale distribution of galaxies (G. Efstathiou, Institute of Astronomy, Cambridge) points to the existence of large-scale coherence. Numerical simulations of galaxy clustering only match the observed correlation function if collapse on a large scale occurs at a relatively late epoch, corresponding to a redshift of less than three. Could galaxy formation coincide with this collapse phase, and have occurred at such a recent epoch? To prove or disprove this hypothesis is the immediate challenge facing the observers of remote galaxies.

---

1. On 'Spectral evolution of galaxies' held at the Rutherford Appleton Laboratory, Oxford, on 23–25 May 1983.

Our Local Pancake

What can be learned about the mechanism of galaxy formation from observations of the Local supercluster, the array of some thousands of galaxies, including those of the Virgo cluster, which is the only one easily accessible to astronomers? The question has been made pointed by the now popular view that only large density fluctuations, containing 10^{14} solar masses and which are therefore perhaps 10^3 times as massive as the Milky Way system, can have survived from the early Universe. Studies of the structure of the Local supercluster (confirming, among other things, that there is indeed a structure) have a bearing on this theory, suggesting that the supercluster may be the evolved state of a minimal surviving density fluctuation.

The prevailing view is that in the early Universe, small-scale structure was suppressed by the diffusive smoothing due to radiation to which matter was coupled or, if massive neutrinos were dominant, by collisionless phase mixing and damping. Only when matter could move freely and when it first combined into atoms, some 3×10^5 years after the Big Bang, was gravitational instability effective at enhancing the growth of density.

Pressure gradients would have been negligible at this stage, and Zel'dovich and his colleagues in Moscow have argued that matter in a region of density fluctuation whose magnitude was of order unity over its extent would probably collapse to what would initially be a highly anisotropic pancake-like configuration. Recent numerical simulations suggest that both filamentary and sheet-like structures may be the generic outcome of such collapses, with great voids separating the highly compressed regions to form an almost cellular structure. The characteristic scale is about 5 Mpc if baryonic matter predominates, but may be some-

Originally published as "Mapping the Local Supercluster," Joseph Silk, *Nature*, 299, 577–578, 14 October 1982.

what larger if massive neutrinos play that role.

On this theory, the dense sheets of compressed gas cool rapidly by radiation and become unstable to further fragmentation. Galaxies form from the fragments, initially in flattened or elongated configurations which are likely to be only transient, because continuing gravitational collapse and infall tends to make the galaxy distribution spherical. However, if the Universe is open, with a present mean density of about one-tenth the critical value for closure, gravitational forces do not significantly modify the pancakes if the initial collapse is fairly recent, say at a redshift less than 10.

To what extent is this theory consistent with what is known of the distribution of galaxies? Recently, data bearing on the theory have come from three-dimensional maps of the galaxy distribution and from redshift surveys. There has been a substantial advance on earlier studies of galaxy counts, which yielded only two-dimensional information. Vast regions devoid of luminous galaxies have been discovered, and the galaxies are often distributed in great chains of clusters and superclusters. This is certainly in qualitative agreement with the pancake theory, but it remains difficult to make quantitative comparisons of predictions with observations.

Studies of the galaxy distribution in our immediate neighbourhood may provide the closest and most detailed view of large-scale structure. It has been known for many years that there is a clear excess of nearby galaxies (above background) in a region that extends over 90° or more near the north galactic pole and called by de Vaucouleurs the Local supercluster. The Local Group of galaxies, of which our own Milky Way galaxy and the Andromeda galaxy Messier 31 are the most prominent, lies on the outskirts of the Local supercluster.

The most detailed chart so far of the Local supercluster has been compiled by R. Brent Tully, who measured some 2,200 galactic redshifts extending to about three times the distance of the Virgo cluster. He finds that the Local supercluster possesses structure and comprises two distinct components—a thin disk containing sixty per cent of the luminous galaxies with the remainder in an inhomogeneous halo. The thickness of the disk is only 2 Mpc (if the Hubble constant is 100 km s^{-1} Mpc^{-1}), whereas its diameter is 12 Mpc. In the halo component, almost all luminous galaxies appear to be in a small number of diffuse clouds, so that most of the halo volume is empty. The more prominent clouds of galaxies are elongated, and their long axes are directed towards the centrally situated Virgo cluster.

To understand this remarkable structure, Tully argues that the disk of the Local supercluster can be so thin only if random motions of galaxies normal to the disk are less than 100 km s^{-1} or if there are very substantial amounts of unseen matter dominating the gravitational attraction within it. The prolate structures found in the halo are suggestive of tidal interactions soon after the halo clouds were formed. The disk thinness is very suggestive of the pancake theory, as is the existence of large volumes devoid of galaxies, but the halo clouds pose more of a problem. The mass of a cloud is only about 10^{13} solar masses—uncomfortably low compared with the predicted minimum pancake mass.

The cloud structures could conceivably arise by fragmentation of pancakes. The details of pancake collapse are sensitive to assumptions about the initial distribution of density fluctuations, and only a modest amount of fragmentation is required. But there are more exotic possibilities if gravitinos or other very weakly interacting particles provide the major contribution to the mass density of the Universe. One needs particles that are even more weakly interacting than neutrinos: they decouple from the Universe at an earlier epoch, when density fluctuations of mass comparable with the horizon size can first grow. In this way, it is possible to have the smallest structures collapse with masses of 10^{13} solar masses or even less.

The redshift survey of the Local supercluster, providing information on the velocity field, is also relevant. Aaronson et al. combine data on galaxy redshifts with measurements of IR luminosities and neutral hydrogen line profiles. The Tully–Fisher relation, which expresses the empirical correlation between galaxy luminosity and hydrogen profile width, was used to infer the distances to all galaxies in a restricted sample. The authors thus have independently derived distances and velocities, and so are able to develop a model for the velocity field in the Local supercluster.

They find that the excess in the local density decelerates galaxies relative to the universal expansion. In addition, the Local Group could have a random component of motion relative to nearby galaxies. The result of fitting the data to the model is a value for the deceleration pattern at the position of the Local Group of about 250 km s^{-1}. This sets an important constraint on the mean cosmological density, which cannot exceed a value of about one-third of the critical density for a closed Universe. If the mean density were any larger, a correspondingly larger gravitational acceleration would be experienced by the Local Group than is inferred from the observations.

One surprising result has emerged from the analysis of the velocity field in the Local supercluster. Aaronson et al. also examined a model which allowed rotation about the supercluster center in addition to deceleration from the Hubble flow. A significant effect was found: the rotation velocity at the position of the Local Group is 180 (±60) km s^{-1}. It may be premature to overemphasize the significance of such a rotation, but it would be a relatively natural outcome of the pancake theory in which coherent formation of large-scale structure occurs. It is also worth noting that the concept of rotation loses much of its meaning for the Local supercluster: the Local Group has completed only a small fraction of a single orbit since the formation of the supercluster.

The pancake theory is now considered a dinosaur of the 1980's. The hot dark matter-dominated Universe, with a neutrino of mass between 30 and 100 electron volts, has been rejected, since it failed to form galaxies sufficiently early. The preferred primordial fluctuation scale does not have a preferred scale of cluster mass: in a cold dark matter-dominated Universe, there is no preferred scale. However numerical N–body simulations of the expanding Universe do demonstrate that transient pancakes form. Today, clusters are collapsing, and indeed, cluster-size pancakes, which subsequently sphericalize, are observed, at least in the computer. The observed Universe as measured in the 3-dimensional galaxy catalogs is a mixture of pancake-like sheets, filaments, and bubbles, of typical dimensions 5—10 megaparsecs.

LARGE-SCALE STRUCTURE

Texture and Cosmic Structure

WITH ROMAN JUSZKIEWICZ

Among the fundamental questions of cosmology are how the original hot primordial soup created in the Big Bang collapsed to become the galaxies we see today and why these are distributed with the particular degree of order that they have. One solution, proposed[1] by Neil Turok in 1989 following an earlier suggestion by R. Davis,[2] was the notion of 'texture', a kind of froth that pervades the Universe and which would have arisen during the first moments following the Big Bang. In a series of papers,[3–7] and colleagues show that texture could produce the observed spectrum of galactic masses and their spatial distribution. Nevertheless, provocative questions remain to be answered concerning this explanation for large-scale structure.

Global texture is a relic of the phase transition that demarcated the end of the grand-unification era, during which the nuclear and electromagnetic forces behaved as different aspects of a single force. An energy field, the Higgs field, was responsible for the grand-unification process when newly created, very massive particles were present that carried the fundamental forces of nature, the nuclear, weak and strong, and the electromagnetic forces. As the Universe cooled below the grand-unification energy scale (about 10^{16} GeV), the uniform part of the Higgs field, occupying most of space, decayed, leaving behind fundamental particles such as quarks and gluons that carry the strong force. However, occasional non-uniformities in the Higgs field survived as topological defects of 'knots' that are infinitesimal ($\sim 10^{-26}$ cm) regions of relic energy density that cannot decay.

Originally published as "Texture and Cosmic Structure," Joseph Silk and Roman Juszkiewicz, *Nature*, 353, 386–388, 3 October 1991.

Defects

Texture is not the only topological defect that can be created by phase transitions in the early Universe. Zero-dimensional, point-like monopoles, one-dimensional strings and two-dimensional walls have all been studied, but none has been shown compellingly to lead to a successful theory of large-scale cosmological structure. But the non-uniform Higgs field represented by texture results in three-dimensional knots with an energy density that generates density perturbations with just the right large-scale distribution to be of interest to cosmologists. Once the horizon scale (the distance travelled by light since creation) becomes comparable to the radius of a texture, the texture collapses owing to the tension associated with the deformation of the Higgs field. As it collapses, it drags in with it all of the associated energy density.

When it reaches a size of about 10^{-30} cm, the knot unwinds itself, emitting a burst of weakly interacting massless particles known as Goldstone bosons. The energy density associated with the texture, and also with the expanding spherical shell of bosons, pushes ordinary matter around, which then seeds the cosmic structure we see today. Immediately after the texture-generating phase transition, the Universe is filled with texture knots of all different sizes which, when the horizon reaches their respective size, collapse. The ambient matter responds to produce density fluctuations that eventually collapse to form galaxies, clusters of galaxies, and structures on even larger scales, leaving behind a background sea of undetectable Goldstone bosons.

There is only one free parameter in the texture model, namely the value of the Higgs field at the moment of symmetry breaking, and which is inferred to be of the order of the grand unification (GUT) scale. This fixes the amplitude of primordial-density fluctuations at first horizon crossing, thereby generating a scale-invariant or 'Harrison–Zel'dovich' spectrum of fluctuations from which large-scale structure eventually develops. A unique feature of the texture model is that the resulting density fluctuations are highly non-random. Normally, theories of structure formation adopt gaussian (or Poisson) initial conditions. With texture, however, there is a greatly increased probability of finding a large-scale fluctuation. This non-gaussian nature of these non-linear objects results in a distinctive pattern on large scales. Rare fluctuations seed large-scale structure and, given the usual normalization to rms fluctuations in the observed galaxy counts on a scale of a few megaparsecs, texture generates more large-scale power than in the standard cold-dark-matter (CDM) cosmological model introduced by Davis et al.[8]

Missing Power

If this feature of the texture-seeded model is real, it will certainly be good news. The problem of the 'missing power' in the standard CDM model has been recognized for several years, since the discovery of large-scale streaming motions[9]. The CDM density fluctuations are too small to explain the fluctuations implied by the streaming[10], as well as those estimated directly from IRAS galaxy counts[11], or implied by the Oxford APM galaxy correlation function[12]. There have been several attempts to save the model and add large-scale power by introducing a cosmological constant[13], or by lowering the biasing factor[14], and hence increasing the initial fluctuations. The texture-seeded CDM approach, seen in this context, is yet another theoretical attempt to add power on large scales while preserving the small-scale successes of the standard model.

Several papers by Turok and collaborators[3–7] develop the implications of the texture model for cosmology. Gooding et al.[3] found that the scaling solution for evolving textures results in a mass spectrum of galaxy halos that declines inversely with the comoving scale of the texture-induced fluctuations. This results in a mass-spectrum of halos proportional to $m^{-1/3}$, in good agreement with the luminosity function of smaller galaxies. Hydrodynamical and dissipative effects in this picture have been investigated by Cen et al.[4] On larger scales, Park et al.[5] have performed N-body simulations in a texture-seeded CDM model, and find galaxy correlations and large-scale streaming motions in reasonable accord with current observations.

Perhaps the most intriguing suggestion was in an article by Turok and Spergel that appeared in *Physical Review Letters*[6]. These authors compute the probability distribution for the relative mass density contrast, δ, in the global-texture model, and explicitly calculate the deviations from a gaussian distribution in terms of the skewness of the distribution. An excess of positive density fluctuations relative to those predicted from a gaussian distribution is found. As in all intrinsically non-gaussian fields, the skewness $<\delta^3>$ in the texture model[6,7] scales like the variance $<\delta^2>$ raised to the power $3/2$. This scaling relation may well provide us with a critical test for distinguishing between texture-seeded models and the conventional picture of gravitational instability. It is of particular interest in view of the positive skewness reported in a recent IRAS-selected sample of galaxies[10].

Does positive skewness imply that our Universe started from non-gaussian initial conditions? Not necessarily. In models with gaussian, small-amplitude primaeval fluctuations, the skewness is zero at early times, and it grows later as a result of nonlinear evolution. This asymmetry in the probability distribution develops because δ can grow indefinitely in regions where the initial $\delta > 0$, whereas in the voids it can never decrease below -1. For weakly nonlinear perturbations, the magnitude of the skewness can be calculated from second-order perturbation theory (to first order, the skewness vanishes). In a flat universe (that is, one with the critical density, $\Omega = 1$, for gravitational closure), the third moment of the density field is $\langle\delta^3\rangle = S\langle\delta^2\rangle^2$, where $S = 34/7$. Since this result was obtained by Peebles in 1980[15] the subject remained dormant, presumably because the existing observational data were too noisy on scales large enough to make perturbation theory applicable. The recent detection of positive skewness in galaxy catalogues has renewed theorists' attention and several groups are now involved in efforts to understand the relationship between different moments of the density field and the time evolution of the probability distribution δ, expected in the gravitational instability picture.

Smoothing

The preliminary results of this research look very promising. To make theoretical predictions comparable to estimates from galaxy redshift surveys, one has to take into account the smoothing of the density field with a filter of a finite width and the mapping from position space to redshift space. Both of these procedures modify S, and the most significant correction, caused by the smoothing, introduces a weak dependence on the effective logarithmic slope $n = \mathrm{dln}P/\mathrm{dln}k$ of the power spectrum P of density fluctuations, measured for wavenumbers k, corresponding to the cutoff induced by the filter. For $\Omega = 1$ and a gaussian filter, the value of S changes from 3.5 to 3.1 when n changes from -1 to 0; and for a top hat filter, $S = 34/7 - 3 - n$. Both the IRAS[16] data and the APM data for $20h^{-1}$ Mpc to $2\pi/k \leq 100\ h^{-1}$ Mpc, where h is the Hubble constant in units of 100 km s^{-1} Mpc^{-1}, appear to be consistent with n of about -1. Moreover, it turns out that gravitationally induced skewness is remarkably insensitive to the exact value of the density parameter: for an open universe[18] ($\Omega < 1$) and infinite resolution the skewness S is enhanced by an infinitesimal additional term, $6/7(\Omega^{0.03} - 1)$.

The weak dependence of S on uncertain parameters like Ω and the shape of the initial spectrum made skewness measurements an excellent tool to test gravitational instability models with gaussian initial conditions (rather than a specific power spectrum). The second and third moments can be (and in fact, have already been) determined from galaxy surveys. Since for small enough δ, S is expected to be constant in the conventional picture whereas the texture model predicts $S \sim \langle\delta^2\rangle^{-1/2}$, the power of the test increases with decreasing variance, and hence our ability to detect departures from gaussian behaviour will improve with the size of the sample. The existing catalogues may actually be already big enough to make the distinction.

1. Turok, N. Phys. Rev. Lett. 63, 2625–2628 (1989).

2. Davis, R. L. Phys. Rev. D35, 3705–3708 (1981).

3. Gooding, A. K., Spergel, D. N. & Turok, N. Astrophys. J. 372, L5–L8 (1991).

4. Cen, R. Y. et al. Astrophys. J. 383, 1–8 (1991).

5. Park, C., Spergel, D. N. & Turok, N. Astrophys. J. 372, L53–L57 (1991)

6. Turok, N. & Spergel, D. N. Phys. Rev. Lett. 66, 3093–3016 (1991).

7. Gooding, A. K. et al. Astrophys. J. 393, 42–58 (1992).

8. Davis, M. et al. Astrophys. J. 292, 371–394 (1985).

9. Dressler, A. Nature 350, 391–397 (1991).

10. Bertschinger, E. & Juszkiewicz, R. Astrophys. J. 334, L59–62 (1988).

11. Saunders, W. et al. Nature 349, 32–38 (1991).

12. Maddox, S. J. et al. Mon. Not. R. astr. Soc. 242, 43P (1990).

13. Efstathiou, G., Sutherland, W. J. & Maddox, S. J. Nature 348, 705–707 (1990).

14. Couchman, H. M. P. & Carlberg, R. G. Astrophys. J. 389, 453–467 (1992).

15. Peebles, P. J. E. The Large-Scale Structure of the Universe (Princeton University Press, 1980).

16. Kaiser, N., in After The First Three Minutes (eds Holt, S. et al.) 248–260 (AIP, 1991).

17. Hamilton, A.J.S., Kumar, P., Lu, E. & Mathews, A. Astrophys. J. 374, L1–L4 (1991).

18. Juszkiewicz, R., Bouchet, F. & Colombi, S. Astrophys. J. 412, L9–L12 (1993).

A Quasar Superstructure

WITH DAVID WEINBERG

There have been occasional reports over the past decade of groupings of quasars, the most luminous and most distant objects in the Universe. Clowes and Campusano, reporting in the March 1991 *Monthly Notices of the Royal Astronomical Society*[1], presented evidence of an elongated grouping, which may both exceed in scale and be far more distant than any known large structure of galaxies.

Clowes and Campusano obtained their data in an automated search of a field of some 25 square degrees, in an area where earlier investigations had suggested an anomaly in the quasar redshift (distance) distribution. An objective prism used with the UK Schmidt telescope allowed low-resolution spectra of many objects to be obtained over a wide area. Quasar candidates were distinguished by a computerized search that looked at both emission lines and excess ultraviolet light. Follow-up spectra were obtained to confirm the quasar identifications and obtain precise redshifts. A total of about 60 quasars were discovered in the field.

By inspecting the redshift distribution, the authors found ten quasars with redshifts between 1.2 and 1.4 (about a quarter of the distance to the most distant known objects). Although there were other peaks in the distribution, statistical analysis suggested that only these ten quasars form a grouping that is likely to be physically associated. The group is elongated, presenting a profile about 1° by 2.5–5°. Its depth probably exceeds $100h^{-1}$ megaparsecs (the exact size depends on the value of Hubble's constant, normally given as $100h$ km s^{-1} Mpc^{-1}, with $0.5 \leq h \leq 1$). Clowes and

Originally published as "A Quasar Superstructure," Joseph Silk and David Weinberg, *Nature*, 350, 272, 28 March 1991.

Campusano found unambiguous evidence for clumping in the field over 1° (or $35h^{-1}$ Mpc), and more tentative evidence for clustering over 5° (in particular, nine of the ten quasars lie above the main diagonal of the survey plate).

There have been several claims for very large structures in the local distribution of galaxies and galaxy clusters. These include the $150h^{-1}$ Mpc 'Great Wall' revealed in the Harvard-Smithsonian Center for Astrophysics redshift survey[2], the $300h^{-1}$ Mpc supercluster complexes identified by Tully[3,4] in the Abell cluster catalogue, and the signals of $120h^{-1}$ Mpc periodicity observed by Broadhurst et al. in deep redshift surveys through the galactic poles[5]. The challenge in such cases is to evaluate the statistical significance of the giant structures, especially because galaxies and clusters are already known to cluster strongly on smaller scales. Postman et al.[6] argue that Tully's results can be explained in terms of the known cluster correlations on scales of about 30 h^{-1} Mpc; and Kaiser and Peacock[7] and Park and Gott[8] show that signatures of 'periodicity' can arise in pencil-beam surveys much like that of Broadhurst et al. as a result of small-scale galaxy clustering, without any special predisposition towards $120h^{-1}$ Mpc structure in the galaxy distribution.

Similarly, although the cold dark matter (CDM) model generates structure only weakly at $150h^{-1}$ Mpc, Park[9] and Gunn and one of us (D.W.)[10,11] find that structures such as the Great Wall arise in numerical simulations of CDM as occasional, chance alignments of three or four smaller superclusters. But statistical studies of the galaxy angular correlation function[12] and the variances of the galaxy density field revealed by the Infrared Astronomical Satellite (IRAS)[13,14] provide clear evidence of an excess of large-scale clustering over that predicted by the standard CDM model.

For the quasar survey by Clowes and Campusano[1], one must ask whether the 100–$200h^{-1}$ Mpc structure is statistically significant, or whether it consists of smaller groups and chance outliers linked by a 'join-the-dots' philosophy. The question is difficult to answer with the current data because the larger scale is close to the angular size of the survey plate itself. An earlier study of a different field by Crampton et al.[15] found a grouping of 23 quasars at a redshift of 1.1, with an inferred physical scale of about $60h^{-1}$ Mpc. Such large quasar groups seem to be rare, because they have not been reported in other faint, wide-angle surveys. Complete surveys of larger areas and greater depths should make it possible to assess the frequency and statistical significance of large quasar groups.

In contrast to the large galaxy structures, rare groupings of rare objects need not be associated with physical increases in the local mass density. If quasars formed from rare peaks of primordial density fluctuations that had random phases, one can readily show that they would have enhanced correlations[16]. The clustering of Abell clusters has been explained in this manner, and if there is typically one quasar formed per galaxy cluster, then similar correlations would result for the quasar distribution. This 'biasing' mechanism might easily give strong quasar correlations out to scales of about $20h^{-1}$ Mpc, but it is difficult to account for $100h^{-1}$ Mpc structure in this way because one would not expect the underlying density fluctuations to be coherent over such large scales.

One cannot automatically dismiss the alternative possibility that the quasar groups represent physical enhancements in the underlying density of the Universe. This is the more radical view, which might be supported by the discovery of associated enhancements in the distribution of galaxies or X-ray-emitting gas. Such structures would be remarkable both for their total size and because, at a redshift of 1.3, they must already have formed when the Universe was less than half of its present age. They are not yet in conflict with other observations—the isotropy of the cosmic microwave background provides the closest constraint—but they would provide a further nail in the coffin of CDM, and they would probably challenge the traditional assumption of gaussian primordial density fluctuations.

1. Clowes, R. G. & Campusano, L. E. Mon. Not. R. astr. Soc. 249, 218–226 (1991).
2. Geller, M. J. & Huchra, J. P. Science 246, 897–903 (1989).
3. Tully, R. B. Astrophys. J. 303, 25–38 (1986).
4. Tully, R. B. Astrophys. J. 323, 1–18 (1987).
5. Broadhurst, T. J., Ellis, R. S., Koo, D. C. & Szalay, A. Nature 343, 726–728 (1990).
6. Postman, M., Spergel, D. N., Sutin, B. & Juszkiewicz, R. Astrophys. J. 346, 588–600 (1990).
7. Kaiser, N. & Peacock, J. A. Astrophys. J. 379, 482–506 (1991).
8. Park, C. & Gott, J. R. Mon. Not. R. Astr. Soc. 249, 288–299 (1991).
9. Park, C. Mon. Not. R. astr. Soc. 242, 59P–61P (1990).
10. Weinberg, D. H. & Gunn, J. E. Astrophys. J. 352, L25–L28 (1990).
11. Weinberg, D. H. & Gunn, J. E. Mon. Not. R. Astr. Soc. 247. 260–286 (1990).
12. Maddox, S. J., Efstathiou, G., Sutherland, W. & Loveday, J. Mon. Not. R. astr. Soc. 242, 43P–47P (1990).
13. Efstathiou, G. et al. Mon. Not. R. astr. Soc. 247, 10P–14P (1990).
14. Saunders, W. et al. Nature 349, 32–38 (1991).
15. Crampton, D., Cowley, A. P. & Hartwick, F. D. A. Astrophys. J. 345, 59–71 (1990).
16. Kaiser, N. Astrophys. J. 284, L9–L12 (1984).

Texture and the Microwave Background

WITH JAMES G. BARTLETT

The formation of galaxies and their large-scale distribution in the Universe has been the focus of intense theoretical research over the years, and remains one of the outstanding problems facing modern cosmology. It is believed that small irregularities, or perturbations, in the mass distribution early in the Universe subsequently grew to form the galaxies and structures we currently see around us. In 1989 Neil Turok produced a model of the origin of these perturbations which involves some rather exotic objects called 'global textures', relics of the grand unification era when the nuclear and electromagnetic forces behaved as different aspects of a single force.

Turok and D. Spergel have predicted the magnitude of the unique signature that textures produce on the cosmic microwave background (CMB), the radiation remnant of creation. The resulting 'hot' spots, actually patches of sky that are hotter than the mean temperature of the CMB by only 1 part in 10^5, should be detectable by the currently operating Cosmic Background Explorer (COBE) satellite. Verification of the model would not only answer some long-standing questions concerning galaxy formation, but also open a window through which the as yet obscure grand unified epoch could be probed.

Global texture is a 'knot' in otherwise empty space consisting of a Higgs field, a concept first introduced to account for the masses of quarks and leptons and to spontaneously break the electroweak symmetry at an

Originally published as "Texture and the Microwave Background," James Bartlett and Joseph Silk, *Physics World*, 3, 24–25, October 1990.

energy of about 100 GeV (a GeV is the energy imparted to an electron exposed to a 10^9 volt potential difference), when the Universe was one-hundredth of a nanosecond old. More generally, other than the electroweak case, Higgs fields appear frequently in modern elementary particle theory. A Higgs field can interact with other particles and carry energy density. For a given spatial and temporal distribution, the Higgs field carries potential energy specified by a potential function $V(\varphi)$, where φ is the Higgs, determined by the underlying theory. This only depends on the value of the field at each point. There is also energy associated with field gradients, due to the (in general) non-uniform distribution, and with any time variation of φ, both referred to as kinetic energy. A global texture is a nonuniform Higgs distribution, for simplicity assumed spherically symmetric, whose energy density arises solely from the field's kinetic energy. It is a three-dimensional topological knot, and is distinct from pointlike field defects (monopoles), one-dimensional defects (strings), or two-dimensional defects (walls), all of which may be produced in phase transitions in the early Universe but have not yet led to any compellingly successful theory for large-scale structure. The energy density of the configuration, and its non-Gaussian distribution of the induced density perturbations, make texture interesting from the viewpoint of cosmology, and especially structure formation.

A phase transition, which lays down the initial distribution of φ randomly in causally disconnected regions, creates the nonuniformity required for the formation of texture in the Universe. Just when this phase transition takes place is determined by the later effects of the texture. For texture to be useful for forming galaxies, the transition must have occurred during the grand unified epoch when the temperature of the Universe was $\sim 10^{16}$ GeV. At any given time in the history of the Universe, the distance light has travelled since creation is referred to as the horizon scale. Once this becomes comparable to the radius of a texture, the texture collapses due to the tension associated with the deformation of the Higgs field. Prior to this, the texture could not have collapsed, otherwise causality would have been violated. As the texture collapses, it drags in with it all of the associated energy density. When it reaches a size of $\sim 10^{-30}$ cm, the knot unwinds itself, emitting a burst of weakly interacting massless particles known as goldstone bosons. The energy density associated with the texture, and also with the expanding spherical shell of bosons, pushes ordinary matter around, which then forms the structure we see today. Immediately after the texture-generating phase transition, the Universe is filled with texture knots of all different sizes which, when the horizon reaches their respective size, collapse. The ambient matter responds to produce density fluctuations that eventually collapse to form galaxies, clusters of galaxies, and structures on even larger scales, leaving behind a background sea of goldstone bosons that appears to be undetectable.

General relativity tells us that gravitational potential wells blueshift light falling into, and redshift light climbing out of, the wells. Thus the energy density of the texture and its associated potential well result in a distortion of the cosmic microwave background that is primarily due to the escape of photons from gravitational potential wells at the epoch in the Universe when the CMB underwent its last scattering with matter, as well as their infall into local potential wells. There are smaller effects due to time-varying potential wells along the line of sight to this last scattering surface, effectively the cosmic 'photosphere', which in the simplest models occurred at a redshift of about 1000, when the Universe was one-thousandth of its present size. The CMB subsequently propagates freely to us. This effect, known as the Sachs–Wolfe effect after the authors who originally calculated its magnitude for arbitrary density perturbations in the Universe, results, for a global texture, in a small frequency shift, which we would interpret as a temperature fluctuation in the CMB towards the direction of the texture. The magnitude depends critically on whether the light passes the center of the texture before or after collapse. If it passes the center before the collapse, the light ray travels through the incoming texture folds as it climbs out of the potential well, thereby being redshifted. Today, we would detect a slightly cooler region of the CMB sky in this direction. Any CMB light ray passing a texture after collapse follows the infalling texture folds into the potential well and is blueshifted. Today we would detect a slightly higher CMB temperature in this direction. The spherical symmetry of the collapsing texture produces a distribution of disc-shaped patches of lower and higher CMB temperature on the sky.

If the model is to create the observed large-scale structure, the maximum amplitude of the temperature fluctuations is calculated to be $\delta T/T \sim 10^{-5}$. Temperature fluctuations of this magnitude should be detectable with the differential microwave radiometer experiment on board the COBE satellite, which is currently generating an all-sky map of the CMB. In contrast to more standard theories of structure formation, the predicted distribution of amplitudes is non-Gaussian, as is clear from the existence of a maximum possible amplitude.

Thus the texture model predicts a unique distribution of fluctuation amplitudes and an unusual disc pattern on the sky, which will enable cosmologists to fully test the theory and its applicability in describing the evolution of our Universe back to an epoch that was a mere 10^{-38} s after the Big Bang.

Is Omega Equal to Unity?

Cosmologists are obsessed with the idea that Ω, the density parameter of the Universe, defined to be the ratio of the mean density to the critical closure value, is precisely equal to unity. Were it not so, the argument goes, the initial conditions of the big bang must have been highly contrived, to within 1 part in 10^{60}, to allow the Universe to expand to its present dimension. Moreover, inflationary cosmology provides a physical reason why Ω should have been so precisely tuned to the value of unity: all curvature, the manifestation of deviations of Ω from unity, was exponentially inflated away during the first 10^{-35} seconds of the cosmic expansion. A new measurement of Ω over a large region of the Universe has been performed by Edwin D. Loh and Earl J. Spillar. Their result is close to 1.

Astronomers have traditionally measured a value for Ω that is much closer to 0.1 than to 1. This has not deterred cosmologists, whose theoretical studies have mostly focused on the implications of a universe with $\Omega = 1$ for large-scale structure and for the nature and detectability of the dark matter. The measurements of Ω that have hitherto been performed sample a fairly local region of space, out to 10 or 20 megaparsecs. (For comparison, the nearest large galaxy, Andromeda, is 0.67 megaparsecs away.) The standard cosmological tests for the curvature of space do sample almost the entire observable Universe, but have hitherto been imprecise in determining Ω because of the uncertain effects of galactic evolution. It is entirely possible that a component of dark matter, invisible to astronomers because it is nonluminous and uniformly distributed over this relatively small length scale, could dominate the mass density of the Universe, thereby allowing Ω to approach unity when the Universe is fairly

Originally published as "Is Omega Equal to Unity?" Joseph Silk, *Nature*, 323, 673–674, 23 October 1986.

sampled over sufficiently large scales.

The Universe has now been probed out to a redshift of about 0.5, or to a scale of order 1,000 megaparsecs by Loh and Spillar, who developed a novel technique for measuring redshifts of a substantial number of galaxies, based on an approach pioneered by R. Kron and D. Koo. The method consists of taking charge-coupled device images of a deep field in six narrowband filters, ranging in wavelength from 4,250 to 9,000 Å. This yields six colors, which provide what is claimed to be a unique photometric measure of the galaxy redshift. The spectral energy distribution of a galaxy varies with wavelength, there being in particular a pronounced spectral break caused by metal absorption near 4,000 Å in the rest frame. At greater distance from us, the redshift increases and the spectroscopic signature marches through the filter sequence. If distant galaxies have intrinsic energy distributions similar to nearby galaxies, this then provides a measure of redshift that is accurate enough to do a cosmological test.

The test consists of measuring the number density of galaxies as a function of redshift. The number density is equal to its euclidean value if $\Omega = 1$ and space is not expanding. The effect of the expansion reduces the number counted in a static space by a factor that measures the contraction of the proper volume element at earlier epochs. Curvature acts to reduce this further, if the Universe is positively curved, and to increase the relative volume element if the Universe is negatively curved. A low-Ω universe corresponds to a cosmological model in which the curvature of space is negative, and one expects the number counts of objects to be enhanced relative to an $\Omega = 1$ model by about 50 percent at a redshift of 0.75 if $\Omega = 0.2$.

By estimating redshifts for 1,000 field galaxies with a median redshift of 0.5, Loh and Spillar argue that they have measured the galaxy number density to better than about 20 percent. Fitting to the luminosity function of galaxies brighter than a specified flux limit enables them to classify the same objects as galaxies at both low and high redshift. Calibration with spectroscopic redshift determinations of galaxies in clusters at $z = 0.4$ suggests that spectral evolution, which could bias their redshift determinations, is unimportant. If they are not systematically losing (or gaining) galaxies at high redshift because of photometric errors or uncertainties, then Loh and Spillar have succeeded in measuring the volume element as a function of redshift. It does not matter whether the mass in the Universe is luminous or dark. This measurement of the geometry of the Universe yields a determination of $\Omega = 0.9^{+0.7}_{-0.5}$ (95 percent confidence limits) and is consistent with the inflationary prediction of an Einstein-de Sitter cosmology ($\Omega = 1$).

This result is especially noteworthy because it is the first seemingly unambiguous astronomical determination of Ω that yields a value of unity on large (~1,000 megaparsecs) scales. Before accepting this as the new cosmological gospel, however, notes of caution have been urged. Redshift errors may be systematically underestimated because the 4,000-Å spectral indicator has only been calibrated against cluster galaxies, whereas this feature is known to be weaker for the field galaxies that dominate the Loh-Spillar sample. There may be systematic errors in photometry that make the smaller, more distant galaxies appear brighter relative to the nearby galaxies, thereby mimicking the effect of high Ω. Dust in distant galaxies may also affect the photometric redshift determinations. These effects need to be critically evaluated, presumably by time-consuming spectroscopic studies, before the $\Omega = 1$ result can be considered definitive. For now, however, it remains a tantalizing possibility: perhaps observational cosmologists (and other astronomers) should pay more heed to theorists.

Following this early study, analysis of large-scale flows has provided more precise evidence that favors a high Ω universe. These probe scales of 10–50 h^{-1}Mpc, where slight but significant deviations from the Hubble expansion are measured. These deviations provide a measure of the dark matter content of the Universe over 50 h^{-1}Mpc, and are consistent with a high value of Ω.

A Bubbly Universe

Perhaps the most spectacular of the large voids in the three-dimensional distribution of galaxies is the Bootes void, a region at least 50 megaparsecs in diameter that contains no luminous galaxies. Until now such voids have seemed interesting oddities that are largely irrelevant to the large-scale structure of the Universe and to the clusters and superclusters of galaxies. But a survey of the large-scale galaxy distribution by Margaret Geller, John Huchra and Valerie de Lapparent reveals that large voids are not the exception but the rule. The implications for our understanding of the origin of the large-scale structure are profound.

The survey is a systematic collection of redshifts of all galaxies of apparent magnitude brighter than 15.5 in a region measuring 6 degrees by 117 degrees on the sky. The redshifts, via Hubble's law, provide a measure of distance from the observer, giving a three-dimensional map of the galaxy distribution in a limited volume of the Universe. Inspection of the map reveals a striking result: large, apparently empty, quasi-spherical voids dominate space, and the galaxies are crammed into thin sheets and ridges between the holes. Similar maps have been presented previously by Jaan Einasto and colleagues, but the reality of this phenomenon was largely discounted because of the inhomogeneity of their data set. The new data thoroughly sample a localized region of the sky, and unambiguously demonstrate the foam-like appearance of the galaxy distribution.

Two notes of caution should be sounded. No correction is made for galaxy-peculiar velocities; these are motions relative to the Hubble flow that are especially large in rich galaxy clusters and will tend to distort the three-dimensional maps that naively translate redshift into distance. This produces the "fingers-of-God" effect: elongations of the galaxy distribu-

Originally published as "Discovering a Bubbly Universe," Joseph Silk, *Nature*, 320, 12–13, 6 March 1986.

tion towards the observer that are not really present appear in the redshift dimension. Another difficulty is that in any magnitude-limited survey, a broad range of nearby galaxies is sampled. However, the most distant galaxies in the sample are exclusively the most luminous ones. If galaxies of different luminosities vary in their spatial distribution, this could also lead to a systematic distortion of the derived maps; for example, the most luminous galaxies, like the highest mountain peaks, may be more clustered than galaxies of average luminosity.

The largest bubbles are 30–50 megaparsecs in diameter. Of note are the well-defined edges of the galaxy layers. How could this situation arise? There are two possibilities: either galaxies formed selectively, at only the edges of the voids; or they moved after formation so as to evacuate large voids.

Consider the former case. A theory that yields beautiful voids was developed by Jeremiah Ostriker and Lennox Cowie, and independently by Satoru Ikeuchi. According to Ostriker and Cowie, an explosion initiated by many supernovae in a newly forming galaxy drives a blast wave that propagates outwards and sweeps up a spherical shell of ambient gas. A hole is evacuated, and the unstable compressed shell fragments to form more galaxies. These, in turn, develop blast waves; a series of bubbles develop that fill most of space, with galaxies present only on shell surfaces.

Explosive galaxy formation is a seductive hypothesis that reproduces the observed foam-like nature of the galaxy distribution. But there are two difficulties that make most cosmologists sceptical of its applicability. One is the plausibility of the mechanism itself—supernovae explode and clear out holes that are tens or, in rare cases, hundreds of parsecs across, but does this phenomenon really work on scales of tens of megaparsecs? Billions of supernovae are presumed to explode coherently, that is to say, over the crossing time of a galaxy of about 10^8 years to yield a vast explosion. A gas-rich protogalaxy could conceivably have quenched such explosions by radiating away much of the kinetic energy of bulk motion needed to form the voids. A more serious difficulty is that even the most optimistic estimates of energy output fail to create sufficiently large voids. Something else is needed.

The missing ingredient is gravity. Density fluctuations were present at the beginning of time, in the earliest instants of the "Big Bang," and gravity amplifies these fluctuations into the large-scale structure of the Universe. Most cosmologists believe that galaxies originated in this manner rather than by explosive amplifications of primeval seeds, which themselves must be attributed to initial conditions. The large-scale structures (notably the clusters and superclusters) and the voids must certainly have formed by the action of gravity in acting over tens or hundreds of megaparsecs. The advantage of this approach is that the physics of a system in which the only significant force is gravity is well understood. It is necessary to use a large computer to perform a numerical simulation, but statistical measures of large-scale clustering, such as the galaxy correlation functions, are known. But to understand the great voids and the sheet-like regions occupied by galaxies it is necessary to go one step further. The concept of biasing the formation of large-scale structure was introduced by Nick Kaiser, and has been applied to galaxy formation. Galaxies are presumed only to form in the rare peaks of an initially gaussian distribution of density fluctuations. Peaks that occur in a potential large-scale cluster acquire a slight boost that enables galaxies to form; those elsewhere fall below the threshold. The biasing hypothesis enhances the large-scale structure that develops as gravitational forces amplify the initial fluctuations. Voids develop, containing hundreds of 'failed' galaxies, and galaxies form only in the dense regions between voids.

Biasing enables simulations of a universe containing cold dark matter at the critical density for closure to be reconciled with the observational determinations of the density parameter of the Universe. The voids are not really void, but contain matter that has failed to become luminous. The dark matter is more uniformly distributed than the luminous matter, and is an inert background that does not respond to most astronomical tests. The dark matter is weakly interacting and clusters on all scales (hence the label cold). It selectively forms galaxies at an early epoch in the rare density peaks. The theory is deficient in two respects: it fails to reproduce the most massive aggregates of galaxies and to account for the clustering of the great galaxy clusters; and the biasing hypothesis lacks any convincing physical explanation.

It could be that neither the explosive amplification scenario nor the biased cold dark matter scenario will account for the foam-like distribution of galaxies that is emerging from the deep surveys. This is no reason for despair; one possibility, for example, is that a scale of 50 megaparsecs or so could have been built into the initial conditions of the Universe. This is not quite as implausible as it seems—in fact, speculations centered around 'hot' dark matter can readily incorporate a scale of this magnitude. Hot dark matter, usually associated with the massive neutrino, has fallen into disfavour because there is too little small-scale power in the initial fluctuations to explain simply how galaxies formed. But the large-scale distribution of matter seems to fall out naturally, rather like a tidal wave sweeping all before it.

One class of theories forms galaxies but has difficulty accounting for the large-scale galaxy distribution, and another class explains large-scale structure but cannot simply form galaxies. Yet another approach is to abandon the assumption that the primordial fluctuations were part of a gaussian distribution. With non-gaussian fluctuations, rare peaks may stand out, whereas the average level of density fluctuations still remains low to conform with the observed isotropy of the cosmic background radiation. The best-studied realization of such a theory invokes vacuum strings left over from a phase transition in the very early Universe. These topological defects create sites around which galaxies accrete, and can also account for the large-scale clustering of the galaxies. It remains to be seen whether such exotic solutions are necessary to meet the challenge of the redshift survey.

The Large-Scale Structure of the Universe

WITH

ALEXANDER S. SZALAY AND YAKOV B. ZEL'DOVICH

Astronomers have long recognized that the distribution of matter on a cosmic scale must somehow bear the imprint of a very early stage in the history of the Universe. A consistent account of that distribution and its evolution must be developed within the context of the Big-Bang theory, since there is almost universal consensus among cosmologists and astrophysicists that the Big Bang provides an empirical framework within which all cosmological issues can be examined. According to the Big-Bang theory, the Universe began as a singular point of infinite density some 10 to 20 billion years ago and pulsed into being in a vast explosion that continues to this day. In the simplest version of the theory, the Universe expands everywhere uniformly from the singular point. The uniformity of that expansion accounts remarkably well for much important observational evidence: Extragalactic matter recedes from our galaxy at a rate that varies smoothly with its distance, and a cold bath of radiation in the microwave region of the electromagnetic spectrum pervades the sky at a temperature that varies over a few angular degrees to less than one part in 30,000. In spite of these successes, there is compelling evidence that the expansion is not precisely uniform. If it were, matter would fail to coalesce and the Universe would become an increas-

---

Originally published as "The Large-Scale Structure of the Universe," Joseph Silk, Alexander Szalay and Yakov Zel'dovich, *Scientific American*, 72–88, October 1983.

ingly rarefied gas of elementary particles. The stars and the galaxies would not exist.

In order to account for structure in the present state of the Universe, the Big-Bang cosmologist must therefore acknowledge some degree of clumpiness early on. Such early inhomogeneities might be smooth and nearly indistinguishable against the homogeneous background; small fluctuations in the curvature of the early Universe would take the form of slight compressions or rarefactions of matter and energy from region to region throughout space. The amplitude of the fluctuations must be large enough (that is, the variation from the average density must be great enough) to grow into the currently observed aggregations of matter in the time since the Universe began; precisely what that amplitude must be, however, is a matter of much theoretical delicacy. If the initial fluctuations were too large, they would cause variations in the temperature of the microwave background radiation that are not observed. Furthermore, the fluctuations must give rise to the structures of relatively special scale that make up the Universe and not to structures of arbitrary size. Stars, galaxies, clusters of galaxies and even superclusters, or clusters of clusters, can now be identified, but at scales of mass larger than that of the supercluster the Universe is fairly uniform.

The convergence of cosmology and the physics of elementary particles has made it possible to satisfy all these requirements without making any strong assumptions about the early state of the Universe. In particular, no appeal is made to any special scales or patterns of mass and energy at the outset of the expansion, and no new forces are invoked. What is assumed is that soon after the beginning of the Big Bang there were small variations in the density of matter and energy everywhere in the Universe. The variations were the result of superposing low-amplitude, sinusoidally varying fluctuations in the density at every possible wavelength, or scale of length; the amplitudes of the fluctuations were distributed randomly, and so the resulting variations in density were random and chaotic. Thereafter, the present structure of the Universe could have evolved according to reasonably well-understood principles of physics.

As the Universe expanded, the random, free streaming of elementary particles in all directions suppressed all the initial fluctuations below a critical size: the only fluctuations that survived were those that compressed or rarefied masses at least 10^{14} to 10^{15} times the mass of our sun. Gravity then caused some of the compressed masses to contract preferentially along one of the three spatial axes. The initial spectrum of fluctuations thereby gave rise to gigantic, irregular clouds of gas that resembled flattened pancakes. Where the pancakes intersected, long, thin filaments of matter took shape. Some of the clouds remained intact; others broke up to form galaxies and clusters of galaxies. The emergence of an appropriate characteristic scale for the fluctuations was first explained by one of us (Silk). The gravitational formation of the thin layers of matter was discovered by another of us (Zel'dovich). We shall refer to this model as the pancake theory.

The pancake theory in its present form is a tale of two objects at the extremes of physical scale. One is an astronomical system large enough to fill 10^{23} cubic light-years of space; the other is the neutrino, a weakly interacting elementary particle that is almost vanishingly small. If the pancake theory is to be confirmed, both objects must be observed and a nonzero mass must be assigned to the neutrino. Since the two masses, if they both exist, span 80 orders of magnitude, extraordinary procedures are needed to measure them from our own vantage of middle dimensions.

Remarkably, the existence of the required scale of the astronomical system has recently been verified, and its mass offers tantalizing evidence that the pancake theory is on the right track. Systematic measurements of distance for several thousand galaxies have been carried out by determining the red shift in their spectra: the displacement of spectral lines toward the long-wavelength end of the electromagnetic spectrum. The red shift is a Doppler effect, caused by the recession of a distant galaxy from our galaxy. The velocity of the recession can be calculated from the red shift according to a simple mathematical formula, and the distance of a galaxy varies directly with its recessional velocity because the Universe is expanding. A redshift measurement combined with the coordinates of a galaxy on the surface of the sky enables the astronomer to fix the galaxy in space. Three-dimensional maps of the distribution of galaxies have thereby been worked out.

The maps show features quite unlike those of most other astronomical objects: the galaxies are concentrated in enormous sheets and filamentary structures whose greatest dimension, roughly 100 million light-years, is an order of magnitude larger than its lesser dimensions. Such a structure can include as many as a million galaxies; its mass is on the order of 10^{16} suns. Moreover, within each structure the galaxies are not evenly distributed: one can distinguish more densely populated clumps and strings, many of them at the intersection of two sheets. Finally, interspersed among the largest structures are huge voids, virtually free of galaxies, that are between 100 and 400 million light-years across. Much of this picture is based on the work of Jaan Einasto of the Tartu Observatory in Estonia.

The detection of a massive neutrino is much more problematic. Several years ago, theoretical physicists assigned to the neutrino a rest mass of zero, but some more recent theories of elementary particles suggest the neutrino does have a small mass. Several kinds of experiment are under way seeking to detect it. In the most direct method the mass can be inferred if certain variations are found in the decay rate of radioactive isotopes. In 1980 Valentin Lubimov, Evgeny Tretyakov and their colleagues at the Institute of Experimental and Theoretical Physics in Moscow measured the decay rate of tritium, the radioactive isotope of hydrogen. At that time they reported results consistent with a small but nonzero neutrino mass, between 14 and 46 electron volts, which is less than one ten-thousandth the mass of the electron. Recently, the same investigators have confirmed their findings and narrowed the limits of error: they now report a neutrino mass of from 20 to 40 electron volts.

Unfortunately, there has been no independent verification of these results, and so there remains no general consensus on the question of neutrino mass. A second kind of experiment, pioneered by Ettore Fiorini of the University of Milan, is based on the rate of a mode of radioactive decay called double-beta decay that is observed in certain isotopes. Fiorini reported that the neutrino mass can be no greater than 10 to 20 electron volts, based on the decay rate of the isotope germanium 76. The method is less direct than the measurement of tritium decay; the results of Fiorini's experiment can be interpreted as a measure of neutrino mass only if it is assumed that the neutrino is its own antiparticle. On the other hand, if the neutrino and the antineutrino are distinct, the double-beta decay of germanium 76 is modified and a value for the neutrino mass cannot be inferred.

A third method of detecting neutrino mass was first proposed by Bruno M. Pontecorvo of the Joint Institute of Nuclear Research at Dubna in the former U.S.S.R. The method exploits the fact that there are three kinds of neutrino: the electron neutrino, the muon neutrino and the tau neutrino. If the three kinds of neutrino have mass, if the three kinds can appear with varying probability and if the difference between the squares of the masses of any two kinds of neutrino is not equal to zero, quantum mechanics implies that the three kinds of neutrino could oscillate, or freely exchange their identities. Since the oscillations would cause the population of one kind of neutrino to vary with time, the oscillations should be detectable as a change in the population of, say, electron neutrinos along the path of a neutrino beam. Several such experiments have been done, first in 1980 by Frederick Reines and his colleagues at the University of

California at Irvine and later by Felix H. Boehm and his colleagues at the California Institute of Technology and by other workers. At this writing no experimental group has reported unambiguous evidence for neutrino oscillations. Unfortunately, the absence of oscillations could merely indicate that the difference between the squares of the masses of two kinds of neutrino is zero; a failure to detect oscillations is therefore consistent with a finite, or nonzero, neutrino mass.

The prevailing attitude among physicists is that the experimental results do not yet warrant a firm conclusion about the mass of the neutrino. Nevertheless, if the evidence for mass is accepted, the cosmological consequences are far-reaching. In the early 1970's, following an early suggestion by Semyon Gershtein of the Serpukhov Institute of Physics in the former U.S.S.R. and one of us (Zel'dovich), György Marx and one of us (Szalay) at Eötvös University suggested that massive neutrinos could make a dominant contribution to the mass and evolution of the Universe as a whole. This suggestion was made concurrently by Ramanath Cowsik and John McClelland of the University of California at Berkeley. A massive neutrino would also lead inevitably to pancakelike structures on large scales. Before we discuss this effect, however, it will be useful to describe an earlier version of the pancake theory, a theory that ultimately failed certain critical observational tests but that has given rise to the more successful theory.

Astrophysicists believe they have a fairly sound understanding of the physical processes that must have taken place after the first few milliseconds of the Big Bang. The energies of particles colliding with one another at that time were no greater than the energies typically achieved in small particle accelerators, and so a picture of the early Universe emerges when one considers a dense fluid of particles whose individual properties are well known.

By far the most abundant particles in the fluid were the photon, the electron and the three species of neutrino; only relatively small numbers of protons and neutrons were left over from annihilations that took place in an earlier epoch. The electrons and the neutrinos remained in close contact throughout most of the first second and were continuously created and annihilated. Frequent collisions among them guaranteed that energy was distributed randomly throughout the fluid; in other words, the particles were maintained in thermal equilibrium. As the Universe expanded, the density of the particles decreased and collisions became less frequent. Because the energy of a photon varies inversely with its wavelength, the average energy of the photons decreased as their wavelengths expanded with the rest of the Universe, and so the Universe began to cool.

Theoretical investigations that have sought to unify the fundamental forces of nature can now peer even farther back than the first millisecond into the history of the Universe. The theories are called grand unified theories because they attempt to understand the electromagnetic force, the weak nuclear force and the strong nuclear force as distinct low-energy manifestations of a single underlying phenomenon. (Gravity, the fourth fundamental force, has not yet been incorporated.) The energy density at which the three forces become indistinct corresponds to the energy density of the Universe only 10^{-35} second after the start of the Big Bang. The early Universe has therefore come to be regarded as a laboratory for testing the predictions of grand unified theories.

One prediction of the theories is that if the density of matter must fluctuate in the early stages of expansion, the density of the photons, or radiation, must fluctuate also. Nevertheless, the ratio of the density of matter to the density of radiation must always remain the same. According to the general theory of relativity, energy and matter are equivalent as a source of gravity and determine the geometry of space-time. A fluctuation in the density of mass and energy therefore causes a fluctuation in the gravitational field, which is equivalent to a fluctuation in the curvature of space-time. The comprehensive theory of such fluctuations in the expanding Universe, treated within the framework of the general theory of relativity, was developed in 1946 by Eugene M. Lifshitz of the Institute for Physical Problems in Moscow.

It seems reasonable to assume that fluctuations must have existed in the early Universe over a wide range of possible scales. We make the assumption primarily for reasons of parsimony: it would seem arbitrary and entirely fortuitous if the initial fluctuations were to single out regions only of, say, galactic scale. There is an upper limit, however, to the size of fluctuation that can be perceived by any observer at a given time. That limit is the spatial horizon of the observer, which is a sphere, centered on the observer, whose radius is equal to the distance light can travel in the time since the start of the Big Bang. In the standard model of the Big Bang, however, the initial expansion of space and time from the singular point creates a universe far greater than the spatial horizon of any single observer. On the other hand, because the expansion of the Universe is thought to be slowing down, increasing amounts of mass come within any observer's horizon. Fluctuations undetectable in the early Universe become detectable later on, because they begin to be encompassed by the observer's ever widening horizon.

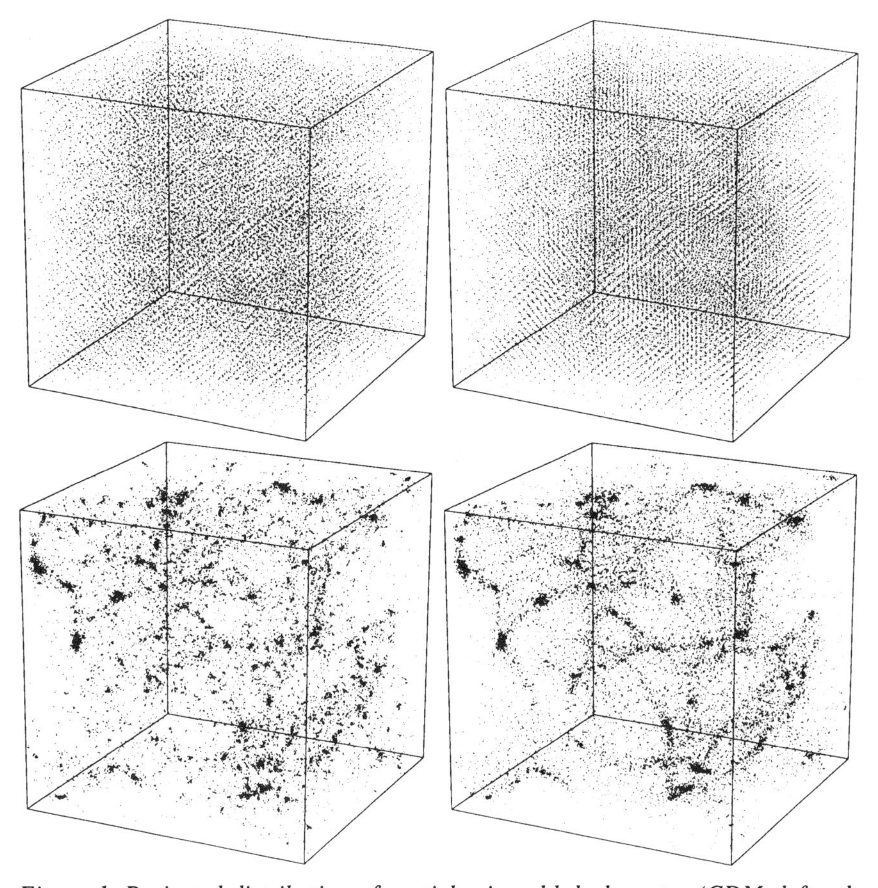

Figure 1. Projected distribution of particles in cold dark matter (CDM; left column) and hot dark matter (HDM; right column) simulations. Shown from top to bottom are the initial conditions, and the state of the system after 10 expansion factors. In each case, the box is 64 Mpc across. These results were obtained by F. Bouchet and L. Hernquist, and are reproduced with their permission.

Once a fluctuation is within an observer's horizon it is adequately described by classical, or nonrelativistic, gravitational theory. There it takes the form of an observable perturbation in the density of the fluid. There are two competing effects on any parcel of matter and radiation moving together: gravity tends to collapse the parcel, and pressure caused by the chaotic motion of the particles and radiation tends to disperse the parcel in space. On large scales gravity always wins. Pressure cannot resist the collapse, so that particles are attracted to the regions of highest density. Moreover, once gravitational collapse begins, the accreted mass can attract more distant mass and radiation, and so any initial instability is amplified. Matter accumulates in some regions and becomes rarefied in others.

If the soup of particles and radiation that made up the early Universe is considered to be an ideal gas, the effects of a density fluctuation superposed on the gas are straightforward. Any local compression in the density over a sufficiently large mass will trigger gravitational instability and lead to incipient collapse. On smaller scales, however, gravity is not strong enough to overcome the increase in the pressure of the gas caused by the increase in density. The compressed parcel of gas will therefore rebound and become rarefied, and the fluctuation will propagate exactly like a sound wave, that is, by the periodic compression and rarefaction of the medium through which it travels.

Most sound waves in the air die out in a few tens of meters because the particles that make up the pressure waves are scattered and their coherent motion is dissipated as heat. In a similar way, sound waves in the cosmic medium set up by the fluctuations lose their energy and die away at all but the longest wavelengths. Furthermore, the particles and photons in the early Universe are much too densely packed to be treated as an ideal gas. In the first 300,000 years of the Big Bang, the photon radiation was energetic enough to keep all matter ionized. Photons outnumbered electrons by a factor of about 100 million; the free electrons, which would later be bound to atomic nuclei, were therefore under constant bombardment by the photons, freely scattering the photons and being scattered by them. The result was a thick, viscous fluid of electrons and photons in which no particle could travel very far without being scattered.

The scattering of the free electrons by the radiation makes any displacement of the electrons through the radiation much like moving through viscous, cold molasses. The viscosity of the fluid thereby inhibits the growth of gravitational instabilities that might be caused by the accretion of matter alone. Moreover, as in the ideal gas, the large pressure of the radiation keeps matter and radiation from collapsing together under the pull of gravity on all but sufficiently large scales. The remaining fluctuations within the viscous fluid, that is, the ones that survive gravitational collapse, can be regarded as sound waves.

As we have mentioned, grand unified theories require that the ratio of the density of photons to the density of matter remain always the same: in the compression phase of a fluctuation the compression of the photons must therefore match the compression of the particles with mass. If the distance covered by a photon in the time since the start of the Big Bang is greater than the distance across a compressed region of a fluctuation, however, the photon will in effect not take part in the compression but will instead dissipate its share of the energy of the fluctuation. Since the photons greatly outnumber the particles with mass, they carry almost all the energy of the fluctuations, and so fluctuations on a scale smaller than the average radial displacement of a photon in the time since the start of the Big Bang are damped out.

The path of a photon can be compared to the path of a drunkard staggering away from a lamppost along any direction with equal probability. In order to wander a distance away from the lamppost that corresponds to N steps when sober, the drunkard must take N^2 steps. Similarly, the photon must be scattered N^2 times in order to travel a radial distance equal to the distance it would travel if it were streaming freely. In spite of their being scattered by the electrons, the photons diffuse radially outward through the medium so fast that they dissipate the energy of all but the largest fluctuations. By the time the Universe cools enough for atomic nuclei to capture the free electrons, the photons have diffused through a region of the Universe whose mass is about 10^{14} times that of our sun. All initial fluctuations that encompass a mass smaller than this amount are therefore erased.

When the electrons finally combine to form atoms, the matter and the radiation cease to interact and the radiation streams off independently of the atoms. The viscosity of the fluid drops abruptly, and the fluctuations that have survived the previous era of radiation-dominated interactions are no longer inhibited from being amplified. Thereafter, gravitational instability proceeds with full vigor.

The sudden absence of radiation pressure has a dominant effect in determining the shape and structure of the first objects to form. Thermal pressure always acts isotropically, or equally in all directions, and so if the radiation pressure had remained comparable in strength to gravity, all collapsed objects would have had nearly perfect spherical symmetry. Anisotropies develop because the pressure is completely negligible up to the last moments of collapse.

Because of the lack of pressure to counteract the infall, gravitational instability is highly efficient at sweeping almost all the matter into compressed, high-density regions of space. Consider the following argument. Along any one of the three spatial axes, matter can be either compressed or rarefied; assume, for the sake of simplicity, that the probability that the matter is compressed along any one axis is one-half. The fraction of the gas that will not be compressed along any axis is the cube of one-half, or one-eighth. This result has immediate implications for the predicted spatial structure following collapse. At an early stage, when the density is still nearly uniform, the regions destined to be compressed include about seven-eighths of all matter. These regions surround smaller bubbles of matter that never collapse; the bubbles are destined to become voids. After the collapse the compressed regions occupy only one-eighth of the volume of space; the small bubbles, which carry one-eighth of the matter, expand to fill the remaining seven-eighths of the volume. The topology of the initial state is preserved. The final outcome is a cellular structure formed by thin walls and filaments of compressed matter that enclose huge voids.

The shape of the compressed regions can be predicted from similar considerations. It is most unlikely that any cubical volume of matter destined for collapse will form a sphere. Such a collapse would require a match of both direction and magnitude of the fluctuations along all three components into which any arbitrary collapse can be resolved. It is much likelier that the cube would collapse first along one randomly selected axis, and it would collapse or expand more slowly along the other two axes. The ensuing distribution of matter is highly anisotropic. Since the mass inside the initial cubical volume does not change as both the thickness and the volume of the cube decrease, the density becomes extremely high and a flat pancake is formed.

At first the pancakes develop in isolated regions, but they soon grow into thin sheets that intersect and form the cellular structure. Numerical simulations of the collapse done with the aid of large computers suggest the Universe has only recently acquired a cellular structure. In the future, as larger clumps of matter form, the cellular structure is expected to disappear. Hence, it is only during an intermediate stage of cosmic evolution that the initial curvature fluctuations are reflected by the structure of matter. The observational evidence shows that from the perspective of the evolution of large-scale structure the Universe is neither very young nor very old.

There are two major difficulties with the pancake theory as we have described it so far. First, remember that the smallest fluctuations to survive the radiation era encompass a mass of 10^{14} suns. Structure in the distribution of galaxies, however, exists at much larger scales. Numerical simulations of a low density universe favor a theory in which the smallest fluctuations emerging from the radiation era are on a scale of 10^{15} or 10^{16} suns.

The second difficulty is more serious. Because the microwave background radiation has propagated freely ever since the photons and the electrons ceased to interact, the variation in the temperature of the radiation across the sky reflects primordial inhomogeneities in the distribution of matter. At the time the original pancake theory was formulated, the upper bound for the temperature variation over the entire sky was about one part in 1,000. Accordingly, it was thought that matter inhomogeneities in the early Universe could have been as great as a third of the temperature variation, or one part in 3,000. But more stringent limits on the variation of the radiation temperature have been set by Francesco Melchiorri and his co-workers at the University of Florence and the University of Rome, and by Yuri N. Parijskiji of the Pulkovo Observatory in Leningrad. The new upper bound is a variation of less than one part in 30,000 over an angle of six degrees.

The fluctuations required by the original version of the pancake theory were consistent with the earlier estimate of temperature variations, but the agreement with the new estimate is only marginal. Moreover, if the overall density of matter and energy in the Universe is so small that the present expansion will continue forever, the agreement is lost. On a cosmic scale the force of gravity would have been so weak in recent epochs that fluctuations must have completed their growth and collapsed at a much earlier time, when the density of matter was much greater than it is today. The amplitude of such fluctuations, however, would have been much too large to be compatible with the uniformity in the microwave background. On the other hand, if the initial fluctuations had been small enough to be compatible with the radiation background, the birth of galaxies would have become practically impossible.

If the Universe is dense enough for the amplitude of the fluctuations to be marginally reconciled with the uniformity of the background, another problem arises. The density cannot then be accounted for solely by the total mass of bright matter, visible as stars, nebulas, galaxies and the like. Instead, the Universe must be made up predominantly of dark matter. This inference is not a new one. Studies of the rotation of our galaxy and that of other spiral galaxies have shown that the rotational velocities of stars on the periphery of a galaxy are not consistent with Kepler's laws. These laws state that the rotational velocity should decrease with increasing distance from the center of a galaxy, just as the orbital velocity of a planet decreases with its distance from the sun. Peripheral stars, however, do not slow down; their rotational velocities are roughly constant and independent of their distance from the galactic center. P. James E. Peebles and Jeremiah P. Ostriker of Princeton University and Einasto simultaneously suggested that the dilemma would be resolved if halos of invisible matter make up the bulk of the mass of spiral galaxies. An indirect argument suggests dark material may be present in even greater quantities within groups and clusters of galaxies. Such systems would fly apart in an unaccountably short time if it were not for the gravitational attraction of dark matter. It is estimated that dark matter may comprise 90 percent of the mass of the Universe.

A new component of the Universe was badly needed to salvage the pancake theory, and a source of dark matter was needed to account for the motions of galaxies. A natural candidate for both purposes was the neutrino, although certain other exotic but still undetected particles, such as a massive photino or a massive gravitino, might serve the same cosmological function. Theories of elementary particles predict that in the first millisecond of the Big Bang a wide variety of weakly interacting particles were maintained in thermal equilibrium. Many such particles could still survive, and provided they are stable they could have far-reaching implications for cosmology. Since the neutrino mass can be measured experimentally, in the remainder of this discussion we shall refer to the neutrino. Nevertheless, even if the neutrino turns out to have no mass, the pancake theory is not disproved.

Remember that in the first second of the Big Bang the primordial soup included an abundance of neutrinos. Even today the ratio of photons to all three varieties of neutrino is only 11 to 9. Neutrinos, unlike protons, electrons and even photons, interact so weakly with other particles that they begin to stream freely through the fluid long before the photons do. Hence the neutrinos, which initially move at the speed of light, can travel farther than the photons in the early stages of the Universe. By the end of the radiation era, the neutrinos have dissipated fluctuations on a larger scale than photons alone could have done.

A massive neutrino cannot continue indefinitely moving at the speed of light. When the energy density of the photons falls below the energy that corresponds approximately to the rest mass of the neutrino, the neutrino begins to slow down and move at a speed appropriate to its energy. If the mass of the neutrino is 30 electron volts, the slowdown will begin well before the capture of the free electrons by atomic nuclei. The capture must wait until the background energy is reduced to 0.1 electron volt, the energy at which hydrogen is ionized by the dense fluid of photons. Although the neutrinos continue to erase the fluctuations as they slow down, they become increasingly susceptible to being trapped by large fluctuations that have not yet been smoothed out. Richard Bond, then at Stanford University, George Efstathiou of Oxford University and two of us (Silk, Szalay) have estimated the maximum scale over which the neutrinos can freely stream before they are trapped, and consequently the minimum scale over which the fluctuations are not erased. The scale corresponds to a present distance of 100 million light-years and a mass of 10^{15} to 10^{16} suns. The agreement with the size and mass of the galaxy superclusters that are now observed is striking.

How can such fluctuations be compatible with the observed uniformity of the background radiation? The neutrinos cease their erasure of the curvature fluctuations before the end of the radiation era, but unlike the electrons their motions are not inhibited by the viscosity of the fluid. Neutrinos collide so rarely with photons or electrons that they are not subjected to viscous drag. Gravitational instabilities among the neutrinos can accordingly begin to develop before the end of the radiation era, and so they can grow over a much longer time than the fluctuations of ordinary matter can. The initial amplitude of the neutrino fluctuations needed to account for the present inhomogeneities of matter could therefore have been much smaller than the amplitude of the fluctuations needed in a mixture of radiation and ordinary matter. With massive neutrinos, the variation in the temperature of the background radiation required to generate the observed aggregations of matter is reduced by an order of magnitude or more. Thus theory and observations can be reconciled.

The new version of the pancake theory leads to a natural explanation for the origin of the dark matter in the Universe. The initial collapse of a pancake distributes most of the neutrinos widely because most of them are accelerated by the collapse to large velocities, on the order of 1,000 kilometers per second. Such neutrinos are destined to fill the dark regions of intergalactic space. Other neutrinos, however, move more slowly because they are initially closer to the central plane of the pancake and do not undergo large accelerations. The thin layer of gas in the vicinity of the central plane condenses and breaks up to form protogalaxies. The slow-moving neutrinos are gathered together by aggregates of ordinary matter, and the matter near the center of the protogalaxy continues to condense and form stars. Neutrinos at the periphery of the protogalaxy, however, are gravitationally shared and never condense; they become the dark matter in the galactic halo.

A more detailed theory of galaxy formation within the context of the new pancake theory is now under development. As a pancake collapses, the neutrino component of the collapsing gas passes through the central plane of the pancake without interaction. The density distribution of the neutrinos acquires sharp discontinuities, some of which can be identified with rich clusters of galaxies. Vladimir I. Arnold, a mathematician at Moscow State University, collaborated with astrophysicists on the problem, and he identified such discontinuities in the overall density distribution with certain elementary structures in the branch of mathematics called catastrophe theory.

The pancake theory, as it has been modified, offers deep insight into the character and origin of the contemporary structure of the Universe. It is based on well-known physical principles and on plausible assumptions about the conditions in the very early Universe. As a theory of the origin of large-scale structure, it is by no means unique, although it appears that both theory and observation point in the general direction we have outlined. Nevertheless, there are many important questions that must be resolved before the theory can be considered firmly established.

Given the confirmation of the theory, there are two main lines along which it must be developed. First, one must address the finer structure of the Universe, the formation of the first generation of stars from a primordial gas that was totally devoid of the heavy elements. Second, one must ask how the initial conditions assumed by the pancake theory arose from even earlier epochs in the history of the Universe. Attempts are now in progress to show how the small-amplitude fluctuations required by the pancake theory could result from earlier and much less well-understood phenomena. These attempts are based on theories not yet settled, but preliminary results leave room for optimism that by the end of the century we shall possess a coherent theory of the Universe.

Recent work by several groups has failed to confirm the measurement of a rest mass for the electron neutrino. The mass is less than 6 electron volts. However, one has ample freedom with either the mu neutrino or the tau neutrino to invoke a rest mass as large as one needs for the resolution of the dark matter problem. Despite the failure of a hot dark matter–dominated Universe to form galaxies in a timely manner, which led to its demise, the role of hot dark matter has recently had a modest revival. With a dominant admixture of cold dark to provide the requisite small-scale power that allows galaxies to form, mixed dark matter (one-third hot, two-thirds cold, equivalent to choosing a neutrino mass of 8 electron volts) works well to account for large-scale structure.

DARK AND LIGHT

The Dark Cloud

The Universe is mostly made of dark practically invisible matter which astronomers have not seen directly but whose existence can be inferred from its gravitational effects. There are well documented cases of galaxy clusters where 90 per cent of the inferred mass is unseen. In individual galaxies, including our own Milky Way Galaxy, substantial amounts of dark matter are present. Cornell University astronomer S. E. Schneider and colleagues report the discovery of a most extraordinary cloud of dark matter—in which there are no detectable stars at all—in the constellation of Leo. This object offers a unique opportunity to study an isolated sample of almost exclusively dark matter.

Ordinarily, in a spiral galaxy such as our own system, stars dominate the inner region. There is also some gas, detected as neutral hydrogen via the 21-cm radio emission line and also via rotational transitions of various molecular species. The gas, like the nearby stars, rotates with respect to the center of our Galaxy. Obscuration by particles of interstellar dust prevent our seeing the more distant stars, but the gas can be studied at 21-cm wavelength throughout our entire Galaxy. The rotation velocity of the gas is found to be relatively constant with increasing distance from the galactic centre. In our own Galaxy, the velocity is about 250 km s^{-1}: the Sun takes some 200 Myr to complete a single orbit. Now, in the Solar System, the rotation speed of more distant planets decreases. We interpret this as simply due to the reduced gravitational pull from the more remote Sun. A space probe sent away from the Sun will decelerate; in contrast, a similar probe moving towards the outer galaxy would not slow down.

The lack of decrease of rotational speed with distance from the galactic centre means that the gravitational potential energy remains approxi-

Originally published as "The Black Cloud," Joseph Silk, *Nature*, 305, 388–389, 29 September 1983.

mately constant. Most of the mass of our Galaxy must be distributed in such a way that the mass M enclosed within a sphere must increase proportionately with the radius R of the sphere. The rotation speed V will then stay roughly constant, since $V^2 = GM/R$, where G is Newton's constant of gravitation.

Within the circle defined by the Sun's orbit around the Galaxy, the total mass enclosed can be readily made up from ordinary stars. This can be seen directly in the ratio of mass to luminosity in the blue spectral region, which takes a value of about 3 solar masses per solar luminosity for the inner sector of our Galaxy. Such a ratio is found for a distribution of stars whose mean mass is about half that of the Sun, as is indeed the case for our Galaxy. A similar situation holds for the visible regions of other spiral galaxies. Only in the outermost regions of a spiral galaxy, where the surface brightness is decreasing rapidly, yet the cumulative mass is inferred from the rotation measurements still to be growing, does one find a ratio of mass to luminosity that is larger than can be interpreted as due to ordinary stars. For the galaxy as a whole, the ratio may be as great as 20 or 30 solar masses per solar luminosity, but of course locally, where there is little light and much mass, it attains even larger values of 100 or more. It is such large values of mass to luminosity that have led astronomers to postulate the existence of exotic objects, such as black holes or very low-mass planet-like stars, in large numbers.

The discovery of the dark cloud in Leo is of considerable importance, for it signifies the first detection of a system containing dark matter and little else. A large isolated cloud of neutral hydrogen was discovered serendipitously during a 21-cm survey of galaxy groups conducted at the Arecibo, Puerto Rico, radio telescope—the largest telescope in the world. There are several groups of galaxies in Leo, but the cloud is not associated with any visible galaxy.

The detection of line emission at the wavelength of 21 cm that characterizes the spin-flip electronic transition in atomic hydrogen allows the velocity of the cloud to be deduced from the wavelength shift relative to the wavelength for emission from atoms at rest. The whole cloud has a systematic velocity away from us of ~960 km s^{-1}. The Hubble law relates galaxy distance to velocity, due to the expansion of the Universe, and from this we can deduce that the cloud is at a distance of 10–20 Mpc.

The total mass of hydrogen gas can now be inferred from the intensity of the 21-cm emission: it is found to be between 8×10^8 and 3×10^9 solar masses, the former value applying if the cloud is at a distance of 10 Mpc. Unfortunately, there is great uncertainty in the estimation of the gas mass. At the low density inferred for the intergalactic cloud, collisional excita-

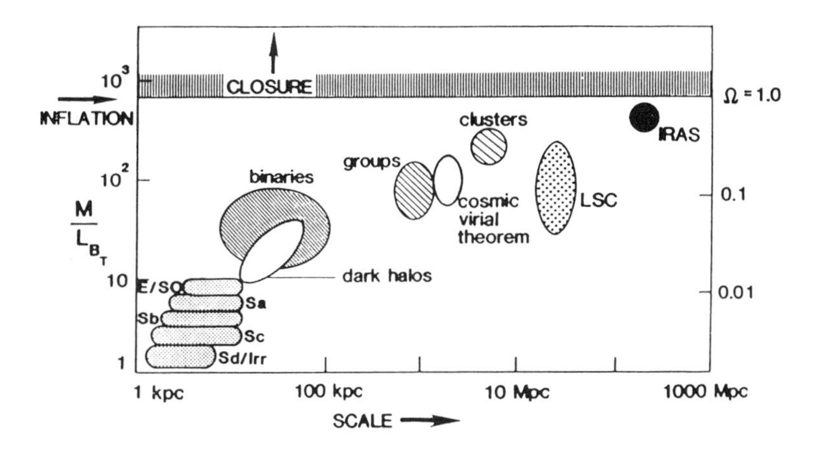

Figure 1. Dark matter in the Universe. The ratio of mass to the luminosity ratio (left scale) is plotted for various systems on scales from a few kpc, as measured for galaxies, up to ~200 Mpc, as inferred from recent observations with the IRAS satellite. The corresponding contribution to the critical density Ω is indicated on the right hand scale.

tions may not fully populate the upper level of the hydrogen atom from which the emission arises. This means that, because of inefficient excitation, as much as an order of magnitude more hydrogen may actually be present than is indicated by the emission line.

The width of the 21-cm emission feature provides the most noteworthy information about the gas cloud, however. It is a measure of the internal velocity field, presumably predominantly due to rotation of the gas, and from it, together with the measured diameter of about 100 kpc, we may infer the total mass. This mass must be present in order for the cloud to be held together by gravity. The inferred gravitational mass is at least 2×10^{10} solar masses, and more likely to be about 10^{11} solar masses. This mass may be as much as 10 times larger than the mass of hydrogen. There appears to be no underlying galaxy, at least with a luminosity of more than about 1 per cent of that of our own Galaxy.

A search on a deep photographic plate of this region by E. Kibblewhite (University of Cambridge) has failed to reveal any galaxy with a brightness in excess of about 27th magnitude (in the red) per square arc second: this sets a lower limit of at least 30 on the ratio of total mass to blue luminosity for the black cloud. This is an intriguing result, for it seems to show that dark matter may accumulate not just in the relatively rare rich clusters, where mass-to-light ratios of 200 or so are often inferred, but also in sparsely populated regions of space. Indeed, the mass-to-light ratio of the black cloud could be infinite, since no underlying galaxy has been detected. A more relevant ratio may be that of dark matter to ordinary matter, in this case hydrogen gas, which is likely to be large: the most probable value is about 5, but a value as large as 100 cannot be ruled out.

In other systems where we have estimates of total mass, ranging from spiral galaxies to galaxy groups and clusters, the ratio of dark to ordinary matter in stars and gas is usually between 3 and 10. The intergalactic cloud in Leo is in accord with this, suggesting that it is likely to be an isolated self-contained sample of the cosmos, with a universal ratio of dark to ordinary matter prevailing over scales from tens to thousands of kiloparsecs. The intergalactic cloud in Leo may provide us with the first isolated sample on galactic dimensions of dark matter without the contamination of stars. The lack of luminous stars is a surprise; a typical spiral galaxy contains a factor of ten more mass in stars than in gas. Somehow, nature seems to have inhibited star formation in the Leo intergalactic cloud.

The potential cosmological significance of this observation is considerable, especially if a large ratio of dark matter to hydrogen is confirmed. For we know that matter in known forms, including stars and gas, falls short of the critical closure density (above which the Universe would be destined to be finite and eventually recollapse, and below which it must be infinite and expand forever) by a factor of, indeed, about 100. If representative of the mean density distribution, the dark matter in the Leo intergalactic cloud provides enough mass to account for about 10 per cent of the closure density.

These are sufficiently dramatic implications that one is well advised to look for any loopholes. The main danger is that the cloud in Leo may not be an isolated gravitationally bound structure. Perhaps the observed gas is held together not by internal gravity but by the pressure of a hotter, more diffuse intergalactic medium. Such intergalactic clouds have been predicted by theories of galaxy formation: hitherto, none as massive as the Leo cloud has been found. The arguments marshalled against this possibility are that, first, no extended X-ray emission characteristic of such a hot medium has been found near the cloud; and second, the velocity structure of the cloud is reasonably regular and the cloud possesses a fairly well defined outer boundary. Such a sharp edge is not expected if the cloud is a transient object that is being dispersed by its own internal motions. The evidence may not yet be overwhelming, but the black cloud in Leo could turn out to provide information of immense significance for cosmology, as well as for theories of galaxy formation.

The Dark Cloud Revisited

WITH CEDRIC LACEY

Intergalactic clouds are believed to be common in the remote Universe, but nearby examples are exceedingly rare. The best example of a large intergalactic cloud that is sufficiently close for there to have been extensive multifrequency studies is in the M96 group of galaxies towards Leo at a distance of about 10 megaparsecs. Stephen Schneider and collaborators[1] have completed a detailed survey of the Leo intergalactic cloud. Their results enable us to discuss whether such a cloud may indeed be a prototype of the intergalactic clouds that are studied at large redshift in absorption towards quasars.

The Leo cloud was originally discovered by Schneider and co-workers[2], by means of its radio emission at 21-cm wavelength from atomic hydrogen (H I). There has since been a possible detection of ionized hydrogen (H II) in the cloud. However, searches for emission of visible light from the cloud have failed to find any evidence for an accompanying population of stars. Searches for emission from interstellar molecules and dust have similarly been negative. The Leo cloud is thus remarkable for an object of its mass in its very low level of star formation activity relative to the amount of gas present. For this same reason, other objects of this type will be difficult to find, but may consitute an important population in the Universe.

Originally published as "The Dark Cloud Revisited," Cedric Lacey and Joseph Silk, *Nature*, 339, 256, 25 May 1989.

Ring Structure

The 21-cm survey, performed at Arecibo and at the Very Large Array, revealed that the H I gas in the Leo intergalactic cloud is distributed nonuniformly around a moderately eccentric ring of diameter 200 kiloparsecs (kpc); for comparison, the diameter of our Galaxy is 30 kpc. The ring's kinematics are those of a keplerian orbit of period 4×10^9 yr, whose center coincides in position and velocity with the centroid of the M105/NGC3384 galaxy pair. Interestingly, the mass-to-light ratio for the pair implied by this is only $25M_\odot/L_\odot$($M_\odot$ and $L_\odot$ are the solar mass and luminosity), comparable to the value one expects for a purely stellar population, suggesting that there is little if any mass in a dark halo around the galaxies, and in contrast to what is found in many spiral galaxies, including our own. The total H I mass amounts to $2 \times 10^9\ M_\odot$, and the mean density is only of the order of 10^{-3} cm^{-3}. However, high-resolution observations[3] reveal that about half the H I is in clumps of size a few kiloparsecs, mass around $10^7\ M_\odot$, velocity width about 20 km s^{-1} and central density 0.1 cm^{-1}. The virial masses of the clumps (deduced from their dynamics) seem to exceed the H I masses by factors of 2–4; if this discrepancy is real, the balance could be supplied by envelopes of H II. The gravitational field of the central galaxy pair will tidally disrupt any gas cloud with a density less than 0.1 cm^{-3}, so the clumps appear to be marginally tidally stable.

A high-sensitivity search has been made for Hα emission in the red spectral region from ionized hydrogen, with a sensitivity to an emission measure of about 1 cm^{-6}pc for gas at 10^4 K, tentative evidence for a detection at this level in the most dense parts of the cloud being reported[4]. The ionization rate required to account for the observed Hα intensity is substantial, amounting to some 6×10^5 hydrogen ionizations per second per square-centimetre column through the clouds. If the H II gas fills most of the volume of the cloud, then its inferred mass exceeds that of the H I. If the gas is photoionized by an extragalactic ultraviolet background, the inferred ionizing flux is only a factor of 2–3 below what one infers by extrapolating from observational upper limits on this background in the 1,400–1,900-Å range[5].

It is also possible that the ionization is produced by the ultraviolet hydrogen Lyman continuum radiation from OB stars, which might be expected to form in the more dense clumps of H I that are comparable in mass and mean density to dwarf galaxies. Sensitive optical observations of one of the most dense clumps show that the maximum surface brightness in the visible-band is 27.7 mag arcsec^{-2}. For comparison, note that for the galactic disk in the solar neighbourhood, the ionizing flux is estimated[6] to be 3×10^7 cm^{-2}s^{-1} for a visible-band surface brightness of 23.5 mag arcsec^{-2} (see de Vaucoulers and Pence[7]; brightness increases with decreasing magnitude). Assuming conservatively that the stars in the clump have the same distribution over mass and age as in the solar neighbourhood, the surface brightness inferred from the ionization rate would be only 27.8 mag arcsec^{-2}, compatible with the quoted upper limit. Given the uncertainties in the stellar mass function and age of the population, it seems that stellar photoionization provides the most natural explanation of the Hα flux.

The disruptive tidal field may provide the reason for the inefficient star formation, as it would inhibit clumping of the gas. The clumps which do exist would be vulnerable to disruption by energy input from young stars if star formation were too vigorous. The distribution of the ring material around a single keplerian orbit suggests that it originated as a single weakly bound object which was then tidally disrupted on close approach to the central galaxy pair—the tidal field being about 10 times stronger at the pericenter of the orbit than at the apocenter. The alternative possibility, that the ring is a shell of ejected gas, can be rejected on account of its negligible radial expansion. The time for the cloud to fall into the system, assuming initial conditions moving with the Hubble expansion of the Universe, would be roughly 6×10^9 yr, so the ring material might only have orbited once in the age of the Universe.

Longevity

The ring owes its longevity to the relatively high angular momentum of the material in it: if it were on a more eccentric orbit, it might by now have been accreted onto the central galaxies. The existence of such material is a natural consequence of the tidal origin of galactic angular momentum in the gravitational instability theory of galaxy formation, in which neighbouring condensations spin up one another. The dispersion in the angular momentum at fixed mass is predicted to be large[8], and should result in some galaxies having reservoirs of outlying gas with high angular momentum.

One expects intergalactic clouds within groups of galaxies to be more common at earlier epochs, because the process of collecting material into galaxies would not have been 100 per cent efficient. This material would only gradually be exhausted as encounters occurred between clouds and galaxies. This late accretion of metal-poor gas could play a role in galactic chemical evolution. It is also tempting to speculate that some of the absorption lines seen in quasar spectra are produced by intergalactic clouds similar to that in Leo. The average H I column density seen in the cloud is a few $10^{18}cm^{-2}$, comparable to that in metal-line absorption systems, and the velocity widths are also comparable[9]. The expected frequency of absorption lines would also turn out to be about right if most galaxy groups at large redshift contained clouds like that in Leo. Perhaps the dark H I cloud in Leo is a prototype of a more extensive population of intergalactic clouds that played a significant role in the early Universe.

1. Schneider, S.E. et al. Astr. J. 97, 666–673 (1989).

2. Schneider, S.E., Helou, G., Salpeter, E.E. & Terzian, Y. Astrophys. J. 273, L1–L5 (1983).

3. Schneider, S.E., Salpeter, E.E. & Terzian, Y. Astr. J. 91, 13–22 (1986).

4. Reynolds, R.J. et al. Astrophys. J. 309, L9–L12 (1986).

5. Martin, C. and Bowyer, S. Astrophys. J. 338, 677–706 (1989).

6. Abbott, D.C. Astrophys. J. 263, 723–735 (1982).

7. de Vaucoulerus, G. & Pence, W.D. Astr. J. 83, 1163–1173 (1978).

8. Barnes, J. & Efstathiou, G. Astrophys. J. 319, 575–600 (1987).

9. Sargent, W.L.W. in Quasar Absorption Lines, (eds Blades, C. & Norman, C.) 1–10 (Cambridge University Press, 1988).

The Invisible Universe

WITH JOHN D. BARROW

For 50 years, galaxies have been seen as the building blocks of the Universe, and have provided our stepping stones to an understanding of its ultimate structure. But the understanding has now progressed to the point where it seems that the bright stars that make up the visible galaxies represent a smaller proportion of the matter in the Universe than the tip of the proverbial iceberg.

It was only in the 1920s that the pioneering work of American astronomer Edwin Hubble established that at least some of the faint patches of light, nebulae, detected with the aid of astronomical telescopes, are not part of our Milky Way system, or even satellites of it. He discovered them to be full galaxies in their own right, comparable in size with our whole Milky Way but so far away that they cannot be seen by the naked eye. The observations led, in 1929, to the discovery that the entire Universe is expanding, carrying these distant galaxies ever further away from us at a rate indicated by the famous red-shift relation, Hubble's Law. Since Hubble's day, the resulting Big-Bang cosmological model has gained impressive observational support because it successfully predicted the existence and temperature of the cosmic microwave background radiation and because it predicts the correct cosmic abundance of helium relative to hydrogen.

As the name Big Bang implies, such mathematical descriptions of the Universe, provided by Einstein's general theory of relativity, start from a state of very high density and expand outwards. The models come in two main varieties: those in which the kinetic energy of the separating galax-

Originally published as "The Invisible Universe," John D. Barrow and Joseph Silk, *New Scientist*, 28–30, 30 August 1984.

ies exceeds the energy of their mutual gravitational attraction, so that the expansion continues forever; and those in which gravity dominates and the attraction is first halted and then reversed. It is a remarkable feature of our Universe that it exists close to the critical dividing line that separates these two radically different long-range forecasts—so close, in fact, that we still cannot say with confidence which future awaits us. The latest cosmological ideas suggest why this should be so.

One of the most interesting ideas to emerge in cosmology, dubbed "inflation", predicts that, as a result of a phase change in the cosmic medium during the first instants of its expansion, the present day Universe, in the post-inflation era, should be expanding at such a rate that it is within one part in a million of that critical dividing line separating models with an infinite future from those that end in a big crunch. Unfortunately, the inflationary scenario does not predict which side of the line we should be. If the Universe does indeed sit as close as this to the critical state, then there must be virtual equality between the energy of expansion and the energy of gravitational attraction—a "coincidence" which itself suggests to some cosmologists an explanation for why the Universe should exist at all. Whatever its origins, however, if the Universe is close to this critical dividing line, then it is a simple matter to calculate the density of matter in the Universe required today to balance precisely (to one part in a million) the observed expansion. This critical density amounts to an average of about 10 atoms per cubic metre throughout the entire Universe. By terrestrial standards such a density is infinitesimal; far smaller than that of the emptiest vacuum we can produce in our laboratories. And yet, the observed density of the Universe is significantly less than this.

If this matter were irregularly distributed, it would be equivalent to finding one galaxy containing a hundred billion stars in every cubic megaparsec of space (one parsec is about 3 • 25 light years). But the typical separation between galaxies with this average mass is seen to be five megaparsecs, not one. So it seems that the luminous matter we now see in stars and galaxies can account for no more than 1 per cent of the critical density predicted by the inflationary cosmology. The inflationary cosmology is so attractive in other ways that this shortfall has provoked a detailed inquiry into where this invisible matter could be hiding, and what form it might take. It leads to further extraordinary possibilities in the world of elementary particles, and by testing these predictions at the particle level it may one day be possible to decide whether the inflationary model provides an accurate description of our Universe.

Protons and ordinary atomic material comprise that part of a galaxy which appears in a photographic image. There does, however, appear to exist another, smoothly distributed, type of cosmic matter that predominates over protons over very large dimensions. In the outermost parts of spiral galaxies, up to 90 per cent of the galaxy's mass may be photographically invisible. The pull of gravity, as measured by the speed at which gas clouds rotate around the galaxy, does not fall off with distance, as does the light distribution. Over larger dimensions, up to tens of millions of light years in clusters of galaxies, the overall gravity field required to hold the cluster together is typically about 10 times larger than that supplied by all the visible galaxies. Therefore, a huge amount of invisible material has to exist somewhere.

Most cosmologists are confident that the missing mass cannot be in the form of ordinary matter, made up of protons and neutrons, which are members of a family of particles called baryons. Studies of the spectral lines in light from old and young stars show up differences in the abundances of helium and deuterium, a fragile isotope of hydrogen that is destroyed, not created, inside stars. These abundances provide compelling evidence that the lightest elements were synthesised from hydrogen during the early stages of the expansion of the Universe, when it was still in a superdense state, less than three minutes after the moment of creation. The rapid expansion of the Universe caused the temperature in this fireball to decrease rapidly, shutting down the nuclear reactions at a time when the helium and deuterium we see in the oldest stars had been made. The rules of this kind of nucleosynthesis are sufficiently well understood for cosmologists to be able to calculate with confidence the proportion of baryonic matter that could have been processed in this way, and thereby to calculate back, from the observed amount of the light elements in the Universe, the total amount of baryonic matter—ordinary matter—that can be present. The permissible range is between 5 and 10 per cent of the critical density. All of this matter could be in the form of ordinary stars without embarrassing the observers too much, although it is quite likely that some exists in the form of hot gas that has not yet condensed into galaxies, or as stars which have a very low mass and are too faint to be seen. But that still leaves 90 per cent of the critical mass unaccounted for.

If the Universe does possess the critical density, this extra mass, all but 10 per cent of its material content, cannot be in the form of ordinary baryonic matter. Its form is completely unknown, except that it is non-luminous. There are many candidates for the dark matter. One of the most widely discussed candidates is the neutrino. Physicists once assumed that the neutrino had no mass at all, but some experiments have recently produced controversial evidence that this elusive particle may have a tiny mass, about 30 electronvolts (the mass of a proton is about a billion eV).

The Big-Bang theory predicts that there ought to be about 100 neutrinos in every cubic centimetre of the Universe today, and even this tiny mass per particle could add up, over such enormous numbers, to dominate the Universe gravitationally. But the experimental results so far are ambiguous. Teams at several different laboratories around the world are attempting to corroborate the claims that the neutrino has mass, and our understanding of the Universe may hinge on the outcome of these experiments. Meanwhile, cosmologists continue to search for other ways to test the hypothesis that the neutrino has mass.

If most of the mass of the Universe is in the form of a sea of neutrinos, then its gravitational influence ought to affect the pattern of galaxies we see in the sky. But the absence of the kind of large-scale structure expected has itself cast doubts on the likelihood of the neutrino possessing mass.

Seeding the Universe

During the first 10,000 years of the universal expansion, conditions would have been so hot that neutrinos moved at almost the speed of light. Such fast-moving particles would rapidly escape from any clouds of higher than average density that might have existed then, unless those clouds were so large that they have expanded, with the expansion of the Universe, to a size a 100 million light years across today. When the Universe cooled and the neutrinos slowed down, these very large neutrino clouds would have remained as the dominant features of the Universe, growing as their gravity attracted other material. So the first structures that formed in the Universe should have been on this scale, which is not the scale of a typical galaxy but rather the scale of a large cluster of galaxies, or a supercluster (a "cluster of clusters"). Although those structures may have started out roughly symmetrical, as they sucked in matter and the clouds of matter collapsed eventually to form galaxies and stars, they would rapidly become asymmetrical. Most likely, one axis would provide a line of least resistance, collapsing the material into a dense sheet, or pancake; a rarer coordinated collapse along two axes would create a thin filament. These sheets and filaments, in this picture, would be regions of high density, protogalaxies, from which galaxies themselves would form by further collapse. And the pancake model predicts that there should be enormous regions of the Universe which contain no galaxies, voids surrounded by enormous sheets of luminous galaxies, and occasional filaments, lines of galaxies where two sheets intersect.

These predictions do bear a superficial resemblance to the observed structure of the Universe. But computer simulations of the way luminous matter clusters in a universe dominated by neutrinos have shown a weakness with the idea. Galaxies form only in the debris after the first "pancakes" collapse, and then move apart from one another, carried along by the universal expansion. Before long, on the universal timescale, the typical dimension of a cluster of galaxies would be more than that of the pancake from which it formed. The theory tells us that the typical structure of a pancake today would be 100 million light years, but the typical scale of galaxy clustering today is less than one-third of this value. That means that, if the model is correct, galaxies must have formed very recently. But if galaxies had formed so recently, we should be able to see many very young, vigorous galaxies, with strong star formation activity still going on. Such objects have not been found, despite intensive searches, and studies of our own Galaxy suggest that it is very old, almost as old as the Universe.

There may be ways to save the neutrino-dominated model, but in the face of these difficulties cosmologists have been lured by particle physicists to consider a more attractive alternative. The upsurge of interest in unified theories of the fundamental forces of nature—the so-called gauge theories—that has accompanied the search for and identification of the W and Z particles has led to predictions that many as yet undetected particles may exist. This particle zoo includes exotic creatures with names like graviton, photino, axion and gluon; most of them are about as massive as the proton, except for the axion which has only 10^{-14} of the proton mass but could be very numerous. None of these particles has yet been detected—nor have other, more exotic, candidates such as magnetic monopoles or primordial black holes. Particle physicists tell us that these species could all exist and be stable, but they cannot predict how many of them there might be in each cubic metre of space today. Such particles are, therefore, all viable candidates for providing the dark matter of the Universe, but none is very attractive unless some independent evidence of its existence is forthcoming. In the absence of hard evidence from the particle accelerators, the cosmologists have turned once again to calculations of the way in which the presence of these exotic particles might affect the large-scale distribution of galaxies.

In a universe dominated by particles of this kind, the irregularities that will grow into the over-dense regions we call galaxies form in much the same way regardless of the specific particle species involved. Unlike the neutrino, the exotic particles have negligible velocities relative to the overall expansion, either because of their mass or, in the case of the axion, because they are formed with negligible motion. So only the smallest clouds of these particles get dispersed during the first few thousand years of the cosmic expansion. The rest survive, and irregularities develop on all scales of size, from clusters of stars to superclusters of galaxies. The smaller structures develop first, and cluster together in a hierarchical fashion, from the bottom up—exactly the opposite of the neutrino pancake scenario. So, can we tell whether the structure in our Universe formed bottom up or bottom down?

Computer simulations of the bottom-up process produce patterns which indicate the expected distribution of galaxies on the sky so long as the total density is five times smaller than the critical value. These are intriguingly suggestive. Groups of galaxies and clusters form, and so do occasional giant filaments and sheets of galaxies. The structure is complicated and diverse, because it starts out from a mixture of irregularities on all scales up to about 100 million light years. Only on scales larger than this does the model show any smooth distribution of mass. Fluctuations of all sizes are condensing at the same time, so there is no problem in understanding why stars as old as the Universe exist in our Galaxy, and the variety of options open to individual fluctuations as they form seems to mirror the variety of galaxies in the real Universe. Real galaxies come in all shapes and sizes, with different rotation rates, chemical compositions and so on, and these variables are found by statistical studies to be both interdependent and dependent on the local density of the region of the Universe a galaxy inhabits. In high-density regions, elliptical galaxies predominate, whereas spirals tend to be found in lower-density environments. The bottom-up scenario provides a natural means of arranging "cross talk" between fluctuations and naturally accounts for these correlations—and the bottom-up scenario is the one that results from the presence of one or more exotic species of particle as the dominant matter in our Universe.

Although the exact identity of the 99 per cent of the Universe we cannot see is still not known, the finger of suspicion points clearly at one family—the gauge particles. The details of this picture of galaxy formation have yet to be fully explored, but its potential richness and complexity, together with the observed richness and complexity of the real Universe, provide the best incentive yet for the particle physicists to try to confirm the reality of some of those exotic particles they speculate about.

From Quark to Cosmos

WITH JOHN D. BARROW

The early Universe has provided particle physicists with an unrivalled accelerator of high-energy particles, while various elementary particles predicted by exotic theories of unification are, to astronomers, increasingly attractive candidates for the "missing" matter in the Universe. The meeting* arranged jointly by the European Southern Observatory and CERN in Geneva was thus a timely forum for particle physicists and cosmologists to exchange ideas, especially because the discovery of the W and Z bosons with masses as predicted by the unified theory of the electromagnetic and weak nuclear forces proposed by Glashow, Salam and Weinberg has renewed confidence in the unified gauge theory route into the elementary particle world.

The simplest grand unification scheme, based on the gauge symmetry SU(5), predicts that protons should decay with a half life of at most 10^{31} yr. E. Fiorini (University of Milan) summarized the status of the continuing search for proton decay. To isolate the rare events due to spontaneous decay of protons, extensive shielding from atmospheric muons produced by cosmic ray showers is required, but, even so, the ultimate sensitivity of these experiments is likely to be limited to about 10^{33} yr by the background neutrino flux, which is high enough to produce an event $\bar{\upsilon}_c + p \rightarrow e^+ n + \pi$, mimicking proton decay with a 10^{33} yr lifetime.

* The meeting on "The large-scale structure of the Universe" took place on 21–25 November 1983.

Originally published as "From Quark to Cosmos," John D. Barrow and Joseph Silk, *Nature,* 308, 13–14, 1 March 1984.

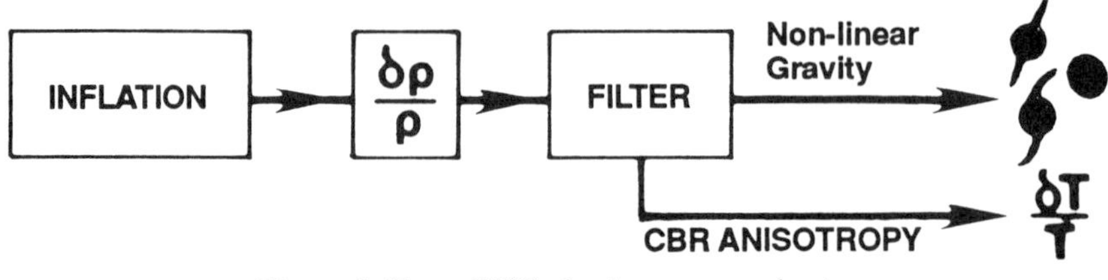

Figure 1. From GUT physics to astrophysics.

Preliminary results were reported at Geneva from experiments carried out deep underground in the Kolar goldfield, Kamioka, the Mont Blanc tunnel, the Silver King mine in Utah and the Morton Salt mine in Ohio. The last experiment provides the most sensitive limit so far, a minimum half life of 1.5×10^{32} yr for the proton. Six candidate events, not due to externally induced reactions, have been found in the other experiments but it is too soon to draw from them definite conclusions about the actual proton lifetime. But the simplest SU(5) scheme of grand unification can now be ruled out. More elaborate theories, predicting proton lifetimes of 10^{32} yr or longer, are readily constructed, however, and have various implications for astrophysics.

One success to report is the possibility of explaining the ratio of protons to photons in the Universe. Since the photons now seen in the 3K background radiation are the remnants of equal numbers of particles and antiparticles created during the thermal equilibrium of the first instants of the Universe, the observed protons represent an excess of matter over antimatter. This asymmetry could have arisen naturally during the first 10^{-35} second (whereafter the grand unification symmetry was broken) provided that protons decayed and charge parity (CP) violation occurred. CP violation has indeed been measured in Kaon decays, but no connection with the early Universe has been demonstrated.

Another expectation of all but the simplest of SU(5) theories and of more exotic unification schemes is that neutrinos should have non-zero mass, although, as the experimenters indicated at Geneva, the theories have not been able to make useful predictions of it. Astronomers have enthusiastically welcomed this development, for even an electron neutrino mass of 10^{-6} eV could, by means of its associated oscillation into other neutrino types, help account for the observed paucity of solar neutrinos. A mass in excess of 1 eV would be significant, since neutrinos would then contribute more than stars to the mass density of the Universe. The Universe would be closed if the neutrino mass were between 25 and 100 eV, the uncertainty arising because of a factor of 2 uncertainty in the Hubble constant, whose square determines the critical closure density. The number of neutrinos is predicted precisely by the Big-Bang theory, although the mass could be spread among several species. Several speakers (P. Darriulat, D. Schramm) pointed out that the existing CERN UA1/2 data on Z° decays already allows an upper bound of about twenty to be placed on the number of neutrino types, although this limit involves assuming a top quark mass of about 30 GeV.

Mössbauer (Munich) reviewed the results of neutrino mass experiments. The only positive result that is not contradicted by other experiments is that of Lyubimov and co-workers in the Soviet Union, which measures the energy distribution of electrons from tritium decay—energy conservation yielding information about the neutrino mass. The experiment is difficult, but the result after several years of data accumulation is that the electron neutrino has a mass of at least 20 eV. Unfortunately, no members of the Soviet group were present to give a detailed account of this crucial experiment.

Of the two other classes of experiment, one measures neutrino oscillations, the conversion of electron neutrinos into mu or tau neutrinos which occurs with unknown probability if either neutrino has a finite mass. Experimental results using targets near nuclear reactors now rule out a mass difference between the two neutrino species of less than about 0.1 eV—provided the mixing probability is not very small. Unfortunately, theorists have no clue as to the likely range of mixing probability, so these experiments are not really conclusive.

If neutrinos are their own antiparticles (that is, are Majorana rather than Dirac particles), the probability of neutrinoless double β decay is greatly enhanced, and T. Kirsten (University of Heidelberg) reported a limit of 6 eV for the electron neutrino mass from studies of the decay of the geochemical abundances of ^{128}Te and ^{130}Te. Similarly, laboratory experiments on ^{76}Ge yield a limit of 14 eV. These limits conflict with Lyubimov's experiment only if the electron neutrino is of Majorana type. Direct limits on the masses of the mu and tau neutrinos are much weaker—0.5 and 250 MeV respectively—and the tau neutrino has yet to be discovered.

At Geneva, these results set the stage for the cosmologists. Speculations are rife about the nature of the dark matter in the Universe, the case for whose pervasiveness was eloquently argued by S. Faber (University of California, Santa Cruz) on the basis of direct determination of the ratios of mass to luminosity in single galaxies, and in groups and clusters of galaxies. A. Sandage (Mount Wilson and Las Campanas Observatories, Pasadena) showed how far we still are from a direct determination of whether the Universe could actually have a closure density in dark matter.

The candidates canvassed range from elementary particles of mass 10^{-5} eV up to supermassive stars and black holes at 10^6 $M_\odot$, allowing some 77 decades of mass for speculation.

Neutrinos with non-zero mass provide at least a verifiable hypothesis, and theories of the large-scale structure of the Universe have now so advanced that observations of the galaxy distribution can be used to test it. The argument went like this:

Inflationary scenarios for the early expansion predict that the mean density is almost exactly at the closure value. Since the primordial synthesis of deuterium and helium in the big bang would conflict with observational evidence if the density in baryons were more than 10 per cent of the closure value (J. Audouze, Institut d'Astrophysique, Paris), the Universe can be at closure density only if non-baryonic matter is sufficiently abundant. Massive neutrinos probably cannot play this part, for reasons that have only emerged in the course of detailed comparisons with the observed large-scale structure of the Universe. J. Oort (University of Leiden), in an account of the very large-scale structure of the Universe, described the enormous structures, extending up to 50 Mpc, that appear in three-dimensional catalogues of the galaxy distribution. Yet, according to D. Wilkinson (Princeton University), the Universe is isotropic and homogeneous to better than 1 part in 10^4 in the distribution of the 3K background radiation when the 180° anisotropy due to our motion is subtracted. Earlier studies nevertheless suggested rather promising agreement in a universe dominated by neutrinos of non-zero mass; a natural scale emerged, determined by the maximum distance neutrinos could stream freely as the universe expanded before they slowed down on account of their mass. Below this scale, roughly matching that of a supercluster of galaxies, no pre-existing fluctuations would have survived, while the first structures to collapse and form galaxies would have had supercluster scale. Ya. B. Zeldovich first showed that highly flattened pancake-like structures would inevitably develop, and fragment into galaxies.

Numerical simulations of non-linear collapse by C. Frenk (University of Sussex) and his co-workers have now revealed a possibly fatal flaw in the neutrino pancake model. The larger the first non-linear scale, the later collapse must occur if the galaxy distribution is to be clustered in the mean to the extent observed. So galaxy formation must have occurred at an unacceptably recent epoch if the Universe is at critical density due to massive neutrinos. Moreover, constraints from the small-scale isotropy of the cosmic background radiation also eliminate any plausible model for a baryon-dominated universe (Silk, University of California, Berkeley).

These arguments are not yet ironclad, but are sufficiently persuasive to direct cosmologists towards other candidates for the dark matter, of which the most attractive is a class of particles collectively labelled "cold relics." Their velocity dispersion has always been negligible over all scales of interest for large-scale structure, which implies that primordial fluctuations, arising for example in the inflationary era, would survive and eventually be gravitationally enhanced. The persistence of small-scale structure, however, implies that galaxies are the first objects to form, but there is expected to be sufficient power on large scales that pancake-like structures should still develop. Inflationary models of the Universe provide the theoretical backing for this expectation: not only can they yield a source of fluctuations but, by producing them on all scales, lead to a unique prediction for large-scale structure.

The candidates for the cold particles include massive photinos and axions. A photino is predicted by supersymmetry to be the fermionic partner of the photon, and is likely to be long-lived. The axion is a coherent field invoked in the theory of strong interactions to circumvent the prediction of a strong CP violation that is in disagreement with experimental limits. In ordinary grand unification schemes, the axion is a boson that, because of its initial coherence, is born essentially at rest when the electroweak symmetry is broken.

Despite their exotic origins, there have been recent proposals to measure the masses of both photinos and axions. Thus D. Sciama (Universities of Oxford and Trieste) pointed out that photinos of mass in excess of about 0.5 GeV would undergo annihilation with antiphotinos, and that while surviving photinos could conceivably be numerous enough to give a critical density, mutual annihilations would result in a possible detectable background flux of γ radiation. As for axions, while the individual mass is only about 10^{-4} eV, there would be about 10^5 axions per photon in the background radiation at closure density, and this large cosmic flux would be further enhanced by axion infall into our local galactic environment. So, in a low-temperature laboratory experiment proposed by P. Sikivie, axions would interact with a strong inhomogeneous magnetic field; those with a mass in the range required to supply the dark material in the halos of spiral galaxies, having a Compton wavelength in the microwave range, might be observable as microwave emission.

Inflationary theories as such did not receive as much attention at Geneva as at previous gatherings of particle physicists and cosmologists. Despite their appealing simplicity—no surviving magnetic monopoles, the present density within one part in a million of the closure value, large-scale homogeneity with a scale-free spectrum of inhomogeneities superimposed—theoretical justification has proved evasive. D. Nanopoulos

(CERN) gave an overview of the inflationary universe model and stressed some of the severe problems that have been recently pointed out. It appears that, at present, inflation remains an appealing idea but that there is no working model that agrees with all the observational constraints.

A highlight of the meeting was the talk by S. Hawking (University of Cambridge) in which he reported recent work performed in collaboration with J. Hartle (University of California, Santa Barbara). It was shown how the quantum state of a spatially closed universe may be described by a wave function obeying a time-independent zero-energy Schrödinger equation. A simple model of an isotropic and homogeneous universe containing a massive scalar field, introduced to represent matter fields in the universe, reveals a wave function for the universe which can be represented as a superposition of different quantum states peaked in probability around particular classical lorentzian solutions of the Einstein equations. These classical solutions are non-singular, begin with a period of exponential expansion and then become matter-dominated before reaching maximum volume and recollapsing. This picture leads to a description of the most probable state for the universe which corresponds, in the classical limit, to one that is singularity-free.

The Missing Mass—Now It's a Gravitino!

Cosmologists have a dire secret. They have not the slightest clue to the nature of the dominant mass constituent of the Universe—only the inherent implausibility of the idea prevents them from postulating the existence of innumerable copies of *Nature* floating in space. Preference has alternated for many years between two candidates: black holes and degenerate black dwarfs of the substellar variety. The mass density in the invisible form must exceed that in luminous material by one to two orders of magnitude.

The amounts of matter and luminosity in a given region are commonly measured in units relative to the mass and luminosity of the Sun. One finds a mass–luminosity ratio of about 2 in the solar vicinity; this means that the matter around us consists of stars typically of somewhat lower mass (luminosity being very sensitive to mass) than the Sun. In a great cluster of galaxies such as Coma, the ratio is about 300 (if one specifies that blue light is measured and that the Hubble constant is 50 km s^{-1} Mpc^{-1}). Part of this difference can be understood in terms of the type of stars present in the cluster. Most of the cluster galaxies are ellipticals, which lack the young, hot, blue stars that contribute little to the mass of a galaxy such as ours but dominate its light. Thus, the mass-luminosity ratio of the underlying old disk and bulge populations of stars near the Sun that are most comparable with stars in ellipticals is about 6. Indeed, this is confirmed by measurements of the velocities of luminous K giants at a height of 1 kpc or more above the disk, which directly sample the local gravitational field. It follows that the total stellar content of the Coma cluster of galaxies amounts to only about 6/300 or 2 per cent of its dynamical mass.

Originally published as "The Missing Mass—Now It's a Gravitino!" Joseph Silk, *Nature*, 297, 102–103, 13 May 1982.

Further, a substantial fraction (about 10 per cent) of the mass of the Coma Cluster has been discovered to be in the form of hot, X-ray-emitting gas. Thus, the total content of the Coma cluster that is in luminous form is only 12 per cent of its total mass. The remaining 88 per cent of the mass of Coma is invisible, at least hitherto, and its nature is a continuing source of speculation. A similar fraction of nonluminous matter appears to be present in the spiral-dominated groups of galaxies (although data on the intergalactic gas content of groups are very incomplete). These measurements both refer to scales of one or two million light years.

On smaller scales, galaxy rotation curves sample the distribution of dark matter in the outer parts of the galaxies. The predominance of flat rotation curves out to several times the optical radius of the galaxy indicates that the mass-to-light ratio is rising with increasing scale. The group and cluster data suggest that the invisible mass fraction of the Universe continues to rise at least up to scales of millions of light years. Whether there is still more dark matter unclustered on larger scales is unknown. One can say that for the Universe to be of critical density, corresponding to closure, the required amount of dark matter has a mass–blue luminosity ratio of about 1,000. Measurements of the redshift–magnitude (or velocity–distance) relation for distant galaxies suggest that the mean density of the cosmos cannot much exceed this critical value. This ratio is only a factor of 3 larger than the Coma cluster value.

Does this signify that the Universe is near its closure density? Almost certainly not, since rich clusters are rather rare objects. Far more typical is the sparse group which has a characteristic mass-blue luminosity ratio of about 50. The large values of mass to luminosity in clusters are due in part to the older (and dimmer) star populations, and in part to the fact that a substantial amount of matter that would perhaps have formed spiral disks is in the intergalactic medium. The phenomenon that is responsible for the dark matter is a universal one. Even studies of galaxy haloes where little light is seen indicate that the ratio of mass to luminosity must locally be extremely large, amounting to several hundred or more. The evidence that exceedingly dark matter dominates the mean mass density of the Universe seems overwhelming. Such matter exists in galaxy clusters and groups, in the haloes of galaxies, and even in the solar vincinity, where, however, the principal contributor to the density is the old star population.

Particle physicists have joined the quest for dark matter in the cosmos. Experimental indications (still unconfirmed) of a finite rest mass for the neutrino have led to a new candidate for the dark matter. The big bang theory predicts that the number density of neutrinos at present should be comparable with the number of photons, both being relics of the primordial fireball that described the early evolution of the Universe. If the neutrino has a rest mass of 1 eV (or one-millionth of an electron mass), it would presently dominate the mass density of the Universe. With a neutrino rest mass as large as 100 eV, the Universe could even be of closure density. The existence of a massive neutrino has notable implications for the large-scale structure of the Universe. Neutrinos are collisionless particles and cannot sustain density fluctuations. As a consequence, all small-scale structure in the Universe would be suppressed, until very late times. This supports a theory of galaxy formation in which the galaxies arise from the fragmentation of large structures that have collapsed anisotropically into pancake-like sheets. Such anisotropies arise naturally in collapse when small-scale structure is suppressed, and lead to the formation of great voids between the clusters and superclusters of galaxies that develop in the high-density regions. One question that remains unanswered in such a theory concerns how the scale of a typical luminous galaxy is determined.

Now a new particle, the gravitino, has emerged from the creative minds of two particle theorists as a new candidate for the dark matter (H. Pagels and J. Primack *Phys. Rev. Lett.* 48, 223;1982). The gravitino is a child of supersymmetry, associated with the search for a theory that unifies gravity along with the strong, weak and electromagnetic interactions. Supersymmetry unites fermions and bosons on a similar basis, and may help explain relations between certain large mass ratios involving the Planck mass, the mass associated with the grand unified theory symmetry breaking, and the mass associated with the electromagnetic–weak force symmetry breaking. Super-symmetry breaks down below energies of about 100 GeV, and results in the production of long-lived fermion partners of the graviton and the photon, the mediators of the only long-range forces. The new particles are called gravitinos and photinos, and both may have finite rest masses. The number of photinos is comparable with the number of neutrinos, and the implications of a finite photino mass are indistinguishable from those of a neutrino rest mass.

However, the gravitino uncouples sufficiently early that the predicted number of gravitinos is only about 10 per cent of the cosmic neutrino flux. Subsequent decouplings of other particles produce the bulk of the neutrinos. This leads to an interesting consequence: if gravitinos are of sufficient mass to account for the dark matter in the Universe, the gravitino rest mass must be some 10 times greater than the neutrino rest mass, or as large as 1,000 eV if gravitinos are at closure density. As the Universe expands, the more massive gravitinos become non-relativistic at an earlier epoch than do the neutrinos. The horizon scale at this instant provides a measure of the minimum primordial fluctuation scale that can survive the free streaming of particles at the speed of light. For neutrinos of mass 30 eV, the perturbation amplitude peaks at 4×10^{15} $M_\odot$, whereas for 1 keV gravitinos the perturbation spectrum peaks at about 10^{12} $M_\odot$. This is just the mass-scale associated with the dark haloes of galaxies. If we take the coincidence seriously, then gravitinos of mass 1 keV are the dominant source of matter in the Universe, and may account for the characteristic scale of galaxies.

Indeed, it turns out, according to calculations by J. R. Bond, A. S. Szalay and M. S. Turner, that gravitinos may lead to the best of all possible scenarios for large-scale evolution. For their mass spectrum extends all the way from scales of supercluster mass down to scales comparable to that of dwarf galaxies. Unlike the neutrinos, gravitinos do not dominate the mass density of the Universe until long after they have become non-relativistic, and this enhances the growth of large-scale gravitino perturbations relative to smaller-scale perturbations, whose growth is inhibited by the inertia of the radiation-dominated mass content of the Universe.

Dark matter may therefore consist of massive neutrinos or gravitinos, or perhaps some more exotic species of elementary particle whose presence arises from unification theories that describe the first 10^{-36} second of the Universe. Such cosminos are the bane of the astronomer's life, for they are likely to be forever invisible, yet they provide a potential solution to some of his greatest problems.

Great Voids in the Universe

Once, astronomers were content to point to the distribution of galaxies in the sky and infer that the Universe is relatively uniform in appearance. More recently, the cosmic microwave background radiation, which samples the Universe at great depth, has provided dramatic confirmation of this homogeneity. It is uniform to at least three-hundredths of a per cent, apart from the dipole anisotropy over 180 degrees that is associated with the motion of our Galaxy. Our local region of space is obviously inhomogeneous, but this can easily be accounted for because the microwave background radiation is seen as being emitted from a very early phase in the history of the cosmos before structure fully developed. One can think of this epoch as corresponding to a cosmic 'surface' or photosphere where the radiation was last scattered, somewhat analogous to the photosphere layer of the Sun below which we cannot see. Before this moment of last scattering, which also coincides with the epoch of 'decoupling' of matter and radiation, the radiation was strongly scattered by ionized particles. As the Universe expanded, the electrons and protons cooled eventually to combine into hydrogen atoms, which are highly transparent to the radiation, and little scattering subsequently occurred.

Since then, small density fluctuations have grown, unimpeded by the drag of the radiation and driven by their own slight increase in gravitational attraction, and have gradually accreted more and more surrounding matter. Such primordial fluctuations eventually formed galaxies, clusters and even superclusters of galaxies. The smoothness of the radiation provides a measure of the degree of inhomogeneity of the Universe at the decoupling epoch, and a strong constraint on theories of galaxy formation.

Originally published as "Great Voids in the Universe," Joseph Silk. *Nature*. 295, 367–368, 4 February 1982.

The larger the structure that is observed, the greater is the challenge to any theory for the origin of large-scale structure. The discovery in 1981 of a great void extending over some 10^6 cubic megaparsecs in the direction of the constellation Boötes therefore attracted considerable attention. Several smaller regions containing no luminous galaxies were discovered during the course of galaxy surveys in the vicinities of great clusters, but the Boötes void is considerably greater than these regions. It had previously been noted that galaxies surrounding the Coma cluster formed a great supercluster. But if galaxies were sprinkled at random locations in space, many of them should have possessed redshifts appropriate to distances intermediate between the Coma cluster and the Virgo supercluster, of which our own Galaxy is an outlying member. Instead, great gaps were found in the redshift distribution, and as redshift is a direct measure of distance, this was indicative of an almost complete absence of bright galaxies in the vast space between Coma and Virgo.

The void in Boötes was discovered by looking deep into three small regions of sky separated by up to 30 degrees, and measuring the redshifts (and hence the distances) of all galaxies brighter than 16th magnitude. In this way, a region of the Universe extending well beyond the Coma cluster was probed. When Robert Kirshner and his colleagues analysed the data in each of the three windows, they found a surprising coincidence—there was a similar gap in the redshift distribution in all three data sets. In effect, they had surveyed three cone-shaped regions that passed through the same great hole in the distribution of galaxies. The only plausible explanation was that the width of the hole spanned at least the 30 degrees they had sampled, and this led to the estimate that it encompassed a roughly cubical volume some 100 Mpc on the side.

The absence of bright galaxies in such a volume can be interpreted in several ways. Perhaps the region is truly devoid of matter, and is a genuine void in space. Galaxy clustering permits the formation of holes, just as in some regions great aggregations develop. This explanation runs into one possible difficulty, since small inhomogeneities should have been present in the matter at the epoch of last scattering of the radiation. The inferred amplitude of the associated radiation fluctuations from which this hole and other similar structures grew must have been at least one part in 10^3, and the fluctuations would extend over an angle of about 2 degrees. One would expect to see traces of these fluctuations in the microwave background radiation; however, recent observations suggest that they are probably absent. These primordial fluctuations could, in principle, be smoothed out if the intergalactic medium stayed ionized because of heating from some unknown process, which would rescatter the radiation.

However, the scale on which such smoothing effects could operate is limited by the distance travelled by light at a time when the density was high enough to permit rescattering of the radiation, and amounts to a few degrees. If the implied radiation fluctuations are definitely not seen, the great void in Boötes cannot be a true hole but must contain matter in some form.

This leads to the intriguing implication that there is an underlying dark component of the matter distribution. One could imagine it as being distributed uniformly on scales of tens of megaparsecs. Only a modest fraction need be clumped to account for the galaxy haloes and clusters which most astronomers believe are bound by such dark matter. It is a sobering reflection on our role in the Universe that most of its contents are hidden from view. The nature of the pervasive dark matter may, however, soon be resolved. Two experiments tentatively reported in 1980 that the neutrino has a finite rest mass. Neutrinos are predicted by the big bang theory to fill space in numbers comparable with those of the photons of the microwave background, and several experiments are underway to confirm this result. If the neutrino mass ultimately is found to exceed even one-millionth of an electron mass, neutrinos would dominate the mass density of the Universe.

Why should the voids contain neutrinos but lack the atoms from which galaxies formed? Adjacent to the voids are huge superclusters of galaxies. Evidently galaxies accumulated in great sheet-like or filamentary structures, occupying a small fraction of the volume of space. The galaxies could not have simply participated in the collapse of the superclusters, otherwise they would have acquired extremely high random velocities that would grossly violate observational constraints. A more plausible possibility is that the matter originally collapsed while still in gaseous form into great sheet-like condensations that subsequently fragmented into galaxies. Gaseous matter can radiate away the kinetic energy acquired during free-fall collapse, and so the newly formed galaxies need not develop large random motions. This idea is the basis of the pancake (or blini) theory of galaxy formation advanced by Ya. B. Zel'dovich and his colleagues in Moscow. A serious but not necessarily fatal objection to the blini theory is that all galaxies are predicted to form at a relatively late epoch. Astronomers have been searching for the luminous protogalaxies predicted by this scenario, but have had little success.

A final possibility is that the galaxy formation process has somehow been modified or even suppressed in the great voids between the luminous galaxies. Perhaps the voids simply lack bright galaxies but contain vast numbers of faint, low luminosity galaxies or, alternatively, enormous amounts of very diffuse hot gas that has failed to condense into galaxies. Existing searches have not been sufficiently sensitive to test either of these hypotheses. One can only conclude, as with so many other astronomical problems, that improved observations are urgently needed.

The Intergalactic Medium

To many astronomers, the search for intergalactic matter resembles the quest for the holy grail. Theory tells us unhesitatingly that the remnants of the huge clouds out of which galaxies formed should still be in intergalactic space. The process of galaxy formation, like practically everything else in astronomy, was inefficient. The vestiges of protogalaxies, or embryonic galaxies, should pervade the Universe as wispy filaments or clouds, and be embedded in a uniform intergalactic medium (IGM) from which every structure we observe today must once have condensed. Theory tells us, and here it develops philosophical undertones, that the Universe was once very simple. A homogeneous and isotropically expanding substrate is the stuff out of which our cosmological models are made, as are the galaxies. Small fluctuations and slight deviations from perfect uniformity were amplified by the inexorable process of gravitational instability, eventually growing into protogalactic gas clouds that collapsed and fragmented into vast systems of stars.

How much matter is left behind is an open question. It is very likely that only a small fraction of the matter in the Universe has formed the visible regions of galaxies. Observations of galaxy rotation and the dynamics of galaxy groups and clusters have indicated the existence of a considerable amount of dark matter. Unseen by any astronomer, this matter nevertheless dominates the gravitational potential in the haloes of galaxies and in bound systems of galaxies. It is a component of the Universe, perhaps the dominant component, that has failed to collapse and make stars in the same manner as the inner regions of galaxies. The dark matter does not consist of gas, otherwise it would be visible. But what it is, we cannot tell. Speculations abound, the current favourite being massive neutrinos.

Originally published as "The Intergalactac Medium," Joseph Silk, *Nature*, 290, 12 March 1981.

The luminous regions of galaxies consist overwhelmingly of stars, and these have formed from clouds of gas. We see this process of star formation continuing today in interstellar molecular clouds. The Milky Way galaxy is a mature system, most of its stars having formed long ago from a much larger gas cloud. The remnants of clouds like this, and the gaseous medium out of which the cloud itself condensed, constitute the elusive IGM.

The search for traces of the IGM has been long and often fruitless. Reputations have been made and broken, often only for what has finally proved to be a new, perhaps stronger, upper limit on what might or might not be present in intergalactic space. Radio, optical, space ultraviolet, X-ray and γ-ray astronomers have all confronted the challenge, and all have failed to find the elusive, uniform IGM.

One early attempt utilized the 21-cm absorption line of atomic hydrogen to search for neutral intergalactic gas along the line of sight to the radio galaxy Cygnus A. No feature was found, and an upper limit was set.

The most stringent limit on the presence of a neutral intergalactic medium came with the discovery of quasars. Sufficiently distant that the cosmological redshift brought the Lyman α line of hydrogen into the visible part of the spectrum, quasar spectra were searched for a possible absorption 'trough' due to intergalactic neutral hydrogen. Failure to detect such a feature meant that, in the region of the Universe that was probed, less than one part in a million of an IGM of density sufficient to close the Universe can exist in neutral form. Subsequent space observations have extended the search for a Lyman α absorption trough into the ultraviolet region of the spectrum in the nearby quasar 3C 273, and confirmed that the IGM, if it exists, must be highly ionized and hotter than 10^6 K (if it is collisionally ionized).

Such a hot IGM would lead to copious X-ray emission, and the discovery in the earliest days of X-ray astronomy of a diffuse background of X-ray emission produced some evidence for it. However, satellite observations with the Einstein observatory revealed that quasars can account for a substantial fraction (over 50 per cent) of the soft X-ray background. At the same time, evidence for hot intergalactic matter has been found in X-ray studies of galaxy clusters. These contain substantial amounts of X ray-emitting gas that is enriched to near the level of the Sun with certain heavy elements. High-resolution observations of quasar spectra have also revealed the presence of discrete intervening clouds of intergalactic gas. But despite the accumulating evidence for intergalactic matter in localized regions, the hypothesized uniform IGM that is so cherished by cosmologists has still not been found.

A 1980 paper by Paul Shapiro and John Bahcall makes an especially intriguing proposal which will extend the search for a Lyman α absorption (which led to a severe limit on a neutral IGM) to the remaining possibility of an ionized IGM. The authors suggest that detection of a hot, ionized IGM may be possible through observations of the X-ray absorption spectra of quasars. Extending earlier calculations by Richard Sherman, they note that quasars, as a class, are sufficiently luminous X-ray sources to be observable at a redshift greater than 2 or 3. If sufficient atoms heavier than hydrogen are present in a hot IGM, they will produce absorption troughs in the X-ray region of the spectrum, due to excitation of inner electrons around the heavy atomic nucleus; iron, for example, has a characteristic absorption at a few keV. Broad absorption troughs in the X-ray spectra of distant quasars would therefore be strong evidence for a hot, ionized IGM with a density amounting to a substantial fraction of the closure density and containing a near solar abundance of heavy atoms. Future space experiments with X-ray detectors capable of moderate energy resolution and high sensitivity should be able to perform this measurement within the next decade.

Dark Matter Comes in from the Cold

Humble baryons, the nuclear material from which our Solar System is made, represent a minuscule fraction, about 4 per cent, of the matter in the Universe. Most of the matter is dark, detectable by its gravitational effects alone. Thus runs the recurrent theme of the Big Bang cosmology to which most astrophysicists subscribe. Observational astronomers, however, urge caution, for solid evidence has remained elusive.

But 1993 has begun with release of new observations that promise to resolve some aspects of the controversies still endemic to the Big Bang cosmology. For the first time, diffuse hot intergalactic gas has been detected, by the ROSAT X-ray satellite, in a small group of galaxies. As described by J. S. Mulchaey, the gas distribution is centered on the group, and the gas temperature is about 10 million degrees Kelvin. Knowledge of the gas temperature with the reasonable assumption that the gas is gravitationally confined allows an estimate of the total mass of the group. Baryons, seen as hot gas and in the luminous galaxies, amount to about 4 per cent (and at most 15 per cent) of the total mass. The remainder is dark matter. For the first time, one can conclusively state that diffuse dark matter dominates the gravitational potential of galaxy groups on megaparsec scales. The more dramatic consequence, however, is that we can now aspire to ascertain the nature of the dark matter.

ROSAT's X-ray scan of the NGC2300 group gives not only the gas temperature but also the abundance of heavy elements, formed in stars long after the Big Bang. The spectrum does not have enough resolution to detect individual elements, but taking the element ratios to be solar, Mul-

Originally published as "Dark Matter Comes in from the Cold," Joseph Silk, *Nature*, 361, 111, 1993.

chaey and colleagues find the abundance of elements heavier than lithium to be about 6 per cent (and no more than 20 per cent) of their abundance in the Sun. Therefore the gas must be predominantly primordial and predates the galaxies.

Nucleosynthesis in the first few minutes of the Big Bang provides a remarkably concordant explanation of the abundances of the light elements: helium, deuterium and lithium. Theory and observation coincide only if the baryon density in the Universe lies within a narrow range: between 4 and 8 per cent of the dark-matter density in a universe at critical density. New evidence that the Universe is indeed at a critical density, destined to expand forever, but at ever decreasing speed, has been presented by K. Kellermann.

Hitherto, astronomers have been able to compare gas and dark matter over large scales only in rich galaxy clusters, where they find diffuse intergalactic gas amounting to at least 10 and even 30 per cent of the dark-matter density in some cases. But that gas is far from pristine, being highly contaminated by ejecta from galaxies—astronomers eagerly await the scheduled launch of the Japanese X-ray satellite ASTRO-D next month [February 1993] to help ascertain the precise location within the clusters of the enriched gas component. Moreover, the complex dynamics of gravitational collapse may operate differentially on gas and dark matter: the X-ray surface brightness and the clumpy distribution of some rich clusters point to this, as does the mapping of dark matter by gravitational lensing. The gas content of rich clusters, even on their outskirts, need not be a good measure of the initial ratio of baryons to dark matter. Only groups of galaxies like NGC2300 can do the trick.

Could the nucleosynthesis prediction be in error? There is of course the cosmological deduction that only a small number (less than four) types of neutrinos are allowed for a unique and simple fit to observations of light-element abundances. The LEP experiments at CERN, as well as results from the Stanford Linear Collider, converged on precisely three types of neutrinos from studies of neutral Z-boson decays. The prediction also relies on the Hubble constant, measuring the expansion of the Universe, and on the temperature of the cosmic microwave background. Only a Hubble constant of about 50 km s^{-1} Mpc^{-1} yields an age for the Universe that is consistent with a universe at the critical density and with the ages of the oldest stars. The direct measures of the Hubble constant (involving supernovae, gravitational lensing, microwave background 'absorption' towards galaxy clusters) all favor such a low value, although indirect techniques mostly indicate a somewhat higher value.

The cosmic microwave background is the remaining ingredient. Its temperature has now been measured to be that of a blackbody with deviations no larger than two-hundredths of one per cent near the peak intensity at a wavelength of 1 mm. The temperature is measured to be 2.726 K, with an uncertainty of only 0.01 K, as reported at Phoenix by John Mather, project scientist for the COBE satellite. Such precision in a cosmological measurement is unprecedented. Combination of the microwave background temperature, the Hubble constant and the observations of the light-element abundances leads to an unambiguous prediction of the baryon fraction. Its measurement in diffuse hot gas that is largely uncontaminated by gas stripped from galaxies points towards a Universe at critical density.

The dark matter is the stuff of particle physicists' dreams, some yet-to-be-discovered form of nonbaryonic matter. Examples abound: from axions, weighing in at 10^{-14} proton masses to supersymmetric relics such as photinos of 100 proton masses or more. We may have to wait until the next millennium for the Superconducting Super Collider and CERN's Large Hadron Collider to suggest anything more definitive. Or perhaps the particles, if they constitute some large fraction of the dark matter of our own galactic halo, will be detected directly.

Nor need seekers of baryonic dark matter be unhappy: the mass detected in luminous baryonic matter—stars and gas—amounts to at most one per cent of the critical cosmological density. The confirmation of a baryon-to-dark-matter fraction of a few per cent, which must await further ROSAT mapping of other galaxy groups, would practically guarantee that the dark halo of our Galaxy is teeming with compact baryonic fragments, most probably aborted stars, brown dwarfs or stellar remnants (old, cold white dwarfs). We can be certain that dark matter is out there, in amounts on the largest scales that totally dominate the dim glimmer of distant galaxies.

LIGHT IN THE DARK

Cosmology Back to the Beginning

Temperature variations in the 3-Kelvin cosmological background, imprinted some 10^{-35} seconds after the initial Big Bang singularity, at the era of inflation, were at last found, 27 years after the microwave background itself was discovered by Penzias and Wilson. At the American Physical Society meeting in Washington, DC, in April 1992 the COBE (Cosmic Background Explorer) team reported detection of the long-sought anisotropies, at a level of only 15 μK. The angular scale of the fluctuations is so large (from 10° to 90° on the sky) that no causal process in standard non-inflationary cosmology could have generated them. We are viewing the birth of the Universe.

All structure in the Universe, according to Big Bang gospel, evolved from infinitesimal density fluctuations. Gravitational instability caused their inexorable growth and eventual development into galaxies, clusters of galaxies, and even larger structures. We observe today structure on scales of up to 100 megaparsecs (Mpc) in the nearby Universe. Yet its origin from density fluctuations has proved difficult to confirm. The reason is simple: such fluctuations were the only deviations from homogeneity in the early Universe.

About a million years after the Big Bang, the density of matter dropped and the radiation temperature cooled to the point at which atomic hydrogen became the predominant state of matter. Before that, the Universe was an ionized plasma in which radiation scattered frequently off electrons, and the two were therefore in intimate thermal contact. But once atomic matter was prevalent, the scattering ceased and radiation was no

Originally published as "Cosmology Back to the Beginning," Joseph Silk, *Nature,* 356, 741–742, 30 April 1992.

longer entrained by matter. Travelling freely from then on, but cooled by cosmic expansion, these photons are now directly observable as the cosmic microwave radiation, which is therefore a pale relic of the fireball that dominated the Universe for the first 10,000 years.

We have two reasons to be reasonably confident in our knowledge of the Big Bang between a few milliseconds and a few years of cosmic time. Synthesis of light elements from hydrogen, particularly helium but also deuterium and lithium, left a remarkably robust legacy of the first few minutes. The Planck spectrum of the cosmic fireball was bequeathed to us after the first few months of cosmic history, when thermalization first became inefficient. The COBE satellite and a rocket experiment have already reported that the cosmic microwave background is a true blackbody at 2.736 K, with deviations no greater than a quarter of a per cent.

These results are the cornerstone of the hot Big Bang cosmology, providing persuasive evidence that the Universe was at one time hot and dense. But a perfectly uniform Big Bang would be unacceptable: there must have been deviations early on to make, over the subsequent billions of years, the obvious structure of the modern Universe. Temperature fluctuations in the microwave background measure the amplitude of these cosmological density fluctuations near the beginning of the era of matter domination, when this process started in earnest.

Until the announcement from the COBE team, there were only upper limits to microwave background fluctuations on various angular scales. These limits were enough to rule out models in which galaxies were made solely from baryonic matter, and constituted a strong rationale for cold dark matter cosmologies. Cold dark matter, weakly interacting particles that are hypothetical relics of the Big Bang, interacts only gravitationally with conventional baryons and radiation, so fluctuations in its distribution can start growing while baryonic matter is still trapped by radiation. When baryons eventually break loose, they fall into the potential wells created by cold dark matter, allowing large-scale structure to arise from much smaller initial baryonic matter fluctuations, with correspondingly smaller variations in the microwave background.

At the transitional epoch when matter and radiation go their separate ways, the horizon scale (how far light travelled since the Big Bang) is naturally imprinted on density fluctuations, and shows up in the distribution of matter today. This scale, about 10 Mpc, is the only natural scale in the cold dark matter spectrum of density fluctuations. Because galaxy clustering is relatively strong on scales of a few megaparsecs, one can normalize theory to observation on a scale where the physics is thought to be simple, and predict, once one has an idea of the initial fluctuation spectrum, what one might expect on much larger scales.

The theoretical breakthrough that came with inflationary cosmology, proposed by Guth in 1980, led to a specific prediction of the shape of the fluctuation spectrum as a function of scale. Inflation, through the presence of a vacuum energy associated with a particle-physics phase transition, causes a period of exponential cosmological expansion; it also blows up tiny quantum fluctuations in spatial curvature to scales that are macroscopic today. The most compelling versions predict equal curvature fluctuations on all scales. This is the famous "n = 1" spectrum, previously advocated on grounds of simplicity by E. Harrison and Ya. B. Zel'dovich. This spectrum is not unique to inflation, but what is remarkable is that inflation predicts fluctuations of this form on scales so large that there has been no subsequent causal contact that could have either generated or destroyed them.

This was the situation on the morning of 23 April 1992. Fluctuations in the fireball radiation, predicted[1–3] by theorists since 1967, had been refined[4,5] to remain just below available experimental upper limits. Their primary cause was gravitational potential fluctuations at last scattering of the radiation, and the scale-invariant spectrum led to the prediction[6,7] of temperature fluctuations of about 1 part in 10^5 over angular scales from 1° to 90°. Above 10°, the prediction refers to acausal fluctuations coming to us directly from inflation; near 1°, one could make a more direct connection to the precursors to structure on 100-Mpc scales. Minimal fluctuations were predicted, on the basis of faith in a gravitational instability origin for galaxy clustering, superclustering and large-scale flows.[8,9]

At the Washington meeting, several groups reported results of their new searches for temperature anisotropies. Observations from the South Pole (University of California, Santa Barbara) and with balloon-borne telescopes (Massachusetts Institute of Technology; University of California, Berkeley) revealed fluctuations at a level of between 1 and 3 parts in 105 over angular scales between 0.5° and 10°. But emission from our Galaxy was seen in the form of low-level fluctuations in diffuse interstellar synchrotron emission, bremsstrahlung and dust emission, and whether any of the temperature variations constituted a residual signal due to the cosmic background remained unresolved.

The Differential Microwave Radiometer (DMR) on board COBE, as reported by G. Smoot (University of California, Berkeley), has important advantages over the rival experiments. Intensity measurements were made at three frequencies, astutely chosen to be both near the galactic minimum and the cosmic maximum intensity (31.5, 53 and 90 GHz), allowing cosmological and galactic emission to be distinguished

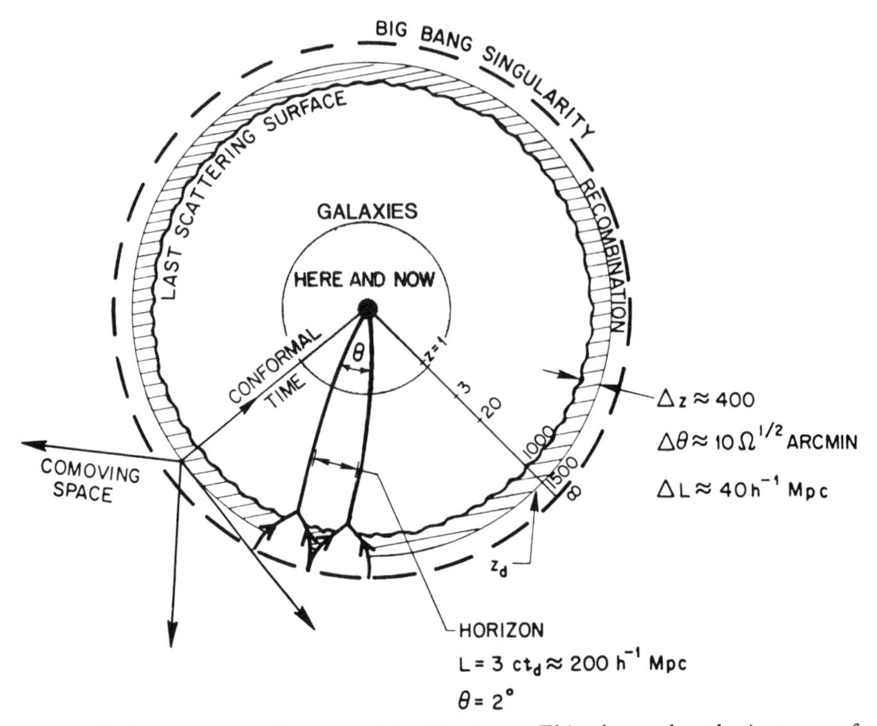

Figure 1. A space-time diagram of the Big Bang. This shows that the isotropy of the cosmic background radiation, when viewed in different directions, does not have a causal origin, at least in the recent Universe. Inflation may provide such an effect, however.

by their different spectral forms.

The DMR has two receiving horns, each equipped with two separate detectors for all three frequencies, pointing 60° apart on the sky, and noise reduction down to a few parts per million is achieved by comparing all the different outputs. The DMR maps the full sky every six months, and with one full year of data analysed, the experimenters unambiguously measured anisotropies in the microwave background. The fluctuations are consistent with the expected blackbody spectrum and have a characteristic amplitude 30±5 μK over 10°–90°. About a tenth of this is due to emission from high latitudes in our Galaxy. This leaves a true cosmological quadrupole signal with amplitude 17±4 μK, or $\Delta T/T = 5(\pm 1.5) \times 10^{-6}$. This is more than a hundred times smaller than the dipole anisotropy due to our motion relative to the cosmic microwave background. The remaining cosmic fluctuations on scales from 7° to 90° have an amplitude of about 6×10^{-6} (in terms of the amplitude of the angular correlation function), and yield an index $n = 1.1 \pm 0.5$, consistent with the scale-invariant prediction of inflationary cosmology.

The observed amplitude indeed matches the density fluctuations required to account for large-scale structure, suitably extrapolated to larger scales according to the scale-invariant prescription. Cold dark matter proponents had even predicted the amplitude, provided one takes the minimal cold dark matter model in a Universe of critical density, without resorting to biasing (peaking) the distribution of luminous matter relative to dark matter. There is a flaw here, worthy of attention: such a model is a disaster on scales of a few megaparsecs, where it predicts excessive gravitational power. Correct the small-scale problem, and cold dark matter fails on large scales.

Other models lurking in the wings bypass inflation entirely, having density less than the critical value, or relying on topological defects such as strings, textures or late-forming domain walls to seed large-scale structure. They also give a quadrupole and low-order multipole structure in the microwave background that is distressingly similar to the scale-invariant, inflationary prediction.

So far, there is no self-consistent set of initial conditions that matches the observed Universe both on large and small scales. Perhaps we should only trust the very large scales where density fluctuations are still small, the nonlinear gravity of galaxy and cluster formation being a complex phenomenon that is still not fully understood. Nevertheless, the detection of fluctuations at more or less the predicted level is a remarkable achievement.

1. Silk, J. Nature 215, 1155–1156 (1967).

2. Sachs, R.K. & Wolfe, A.M. Astrophys. J. 147, 73–90 (1967).

3. Sunyaev, R.A. & Zel'dovich, Ya. B. Astrophys. Sp. Sci. 7, 1–19 (1970).

4. Peebles, P.J.E. & Yu, J.T. Astrophys. J. 162, 816–836 (1970).

5. Wilson, M. L. & Silk, J. Astrophys. J. 243, 14–25 (1981).

6. Bond, J. R. & Efstathiou, G.E. Astrophys. J. 285. L45–L49 (1984).

7. Vittorio, N. & Silk, J. Astrophys. J. 285, L39–L44 (1984).

8. Juszkiewicz, R., Gorski, K. & Silk, J. Astrophys. J. 323, L1–L4 (1987).

9. Gorski, K. Astrophys. J. 370, L5–L9 (1991).

Probing the Primeval Fireball

Radiation from the very farthest depths of the Universe bathes us in a perpetual glow. This emission is only a feeble glimmer from the past, and an isolated thermometer immersed in it would only register a scant 3° Kelvin. Yet once, some 15 billion years ago, this radiation was a white-hot fireball capable of vaporizing any known material. The existence of this so-called cosmic background radiation was predicted more than 40 years ago and was discovered as a hiss of radio static in 1965. It is still one of the strongest pieces of evidence that the Universe experienced a fiery, dense beginning called the Big Bang.

Thanks to a remarkable satellite launched late in 1989 called the Cosmic Background Explorer (COBE), we now know that the spectrum of these radio photons from the depths of space is indistinguishable from that of a blackbody—a perfect absorber and emitter of radiation. Even though the cosmic background retains only a faint echo of its past glory, it still has the characteristic energy distribution of radiation in complete thermal equilibrium with its surroundings. This is the unambiguous signature of the primordial fireball.

Spectral Distortions

Prior to the COBE team's announcement, a number of experiments sought spectral irregularities in the cosmic background. In fact, several groups had reported evidence for deviations from a blackbody spectrum. Such distortions, if real, would be of considerable interest. Departures from an

Originally published as "Probing the Primeval Fireball," Joseph Silk, *Sky and Telescope,* 74, 600–603, June 1990.

ideal radiation field would indicate unknown objects or processes releasing or absorbing energy in the very early Universe.

Think of the early Big Bang fireball as a potent furnace. Its blackbody radiation tells us nothing about what is being consumed. If, however, the furnace is insufficiently hot, "combustion" may not be completely efficient, and some trace of the "fuel" will remain.

As the Universe expands and cools, the processes of photon production and destruction that maintain a perfect blackbody radiation field eventually become too slow to maintain thermal equilibrium. This could occur for the first time about one year after the Big Bang. From then on, traces of the "fuel" may show up as deviations from a blackbody energy distribution imprinted on the spectrum of the background radiation. Indeed, once generated, such distortions are indestructible. They would still be there some 300,000 years after the Big Bang, when the Universe became transparent and the photons of background radiation that we now see began their long journeys.

Even though COBE's observations have ruled out the existence of major spectral deviations, more subtle effects may be present. Departures from a blackbody spectrum might be the fingerprints of energetic phenomena that may once have played an important role in cosmic evolution but have long since faded into oblivion. For example, these departures could tell us about young galaxies burning brightly in their first vigorous phase of star formation. Enshrouded within a womb of dust—the condensed ejecta from the first generation of stars—these protogalaxies emit most of their energy in the far infrared. This emission is redshifted by universal expansion, until, today, it would appear as excess emission at submillimeter wavelengths in the cosmic background spectrum.

A particularly important fingerprint was first postulated by Soviet cosmologists Rashid Sunyaev and Yakov Zel'dovich in 1970. As radio photons pass through hot, dense gas, their energies are boosted by collisions with fast-moving electrons. As photons on the long-wavelength (low-energy) side of the blackbody peak are shifted to the short-wavelength (high-energy) side, they cause a characteristic distortion in the shape of the blackbody curve. If seen, the Sunyaev-Zel'dovich effect would tell us much about the existence of hot intergalactic matter, heated perhaps by the explosive outpourings from giant radio galaxies and quasars. In fact, small-scale examples of this effect have already been detected toward several clusters of galaxies.

Another interesting distortion could arise because the early Universe was an unrivaled creator of massive and highly unstable exotic elementary particles. If some did not decay until the cosmos was at least one year old, the energy they emitted would leave a trace distortion in the cosmic background. In effect, the background radiation is a cosmic wastebasket, collecting the debris from all physical processes occurring since the first year of cosmic history.

Accidental Discovery

Measurements at many different wavelengths are necessary to check the blackbody nature of the cosmic background radiation, and to search for possible spectral distortions. This is no easy task. Indeed, the first reports of such deviations did not occur until four years after the detection of the background radiation by Arno Penzias and Robert Wilson at Bell Laboratories in 1965.

Ironically, their discovery, for which they were awarded the 1978 Nobel Prize in physics, was not motivated by cosmology but rather by a study of diffuse radio emission from our galaxy. In fact, the cosmic microwave background was almost discovered on several previous occasions. Its existence was anticipated by George Gamow in the 1940's, predicted by Ralph Alpher and Robert Herman in 1948, and independently re-predicted over the next 20 years by Robert Dicke in the United States and various other theorists in the Soviet Union, the United Kingdom, and elsewhere.

The story of the cosmic background radiation really begins with the discovery of the cyanogen (CN) radical in interstellar space by Theodore Dunham and Walter Adams in 1937. This species was seen as an absorption line against the star Zeta Ophiuchi and appeared appreciably excited above its lowest-energy or ground state. Four years later, Andrew McKellar calculated that the excitation temperature was 2.3° K, and could be due to absorption of microwaves with this characteristic temperature.

Why then, did Dunham, Adams, and McKellar not discover the microwave background and win Nobel Prizes instead of Penzias and Wilson? The discovery was missed because, at that time, it was thought more likely that the CN excitation was due to collisions with electrons. In yet another irony of history, the CN "thermometer" has since given some of the most precise pre-COBE measurements of the temperature of the cosmic background radiation.

An independent upper limit on the temperature of diffuse microwave emissions was set in 1946 by Dicke and collaborators. Their measurements, made at three wavelengths near 1 cm with a radiometer on the roof of MIT's Lincoln Laboratory, set a limit of 20° Kelvin. Unfortunately, this result went largely unnoticed and was generally forgotten, even by Dicke himself!

The cosmic background radiation was again almost discovered in 1960 by Edward A. Ohm of Bell Laboratories when he was carefully testing and calibrating the horn antenna to be used for the first trans-Atlantic television broadcast via the Telstar satellite. The noise he measured at a wavelength of 10 cm when looking at the sky was 3.3° Kelvin hotter than could be accounted for by a careful evaluation of all spurious sources of emission. Unfortunately the sum of all his uncertainties was about 3° Kelvin, and so his measured excess was not statistically significant. In another accident of history, Penzias and Wilson were using this same horn when they detected the microwave background four years later.

Pre-COBE Observations

Measuring such a weak and diffuse signal as the background radiation poses a considerable challenge to radio astronomers. Spurious emissions from the atmosphere and the ground must be carefully measured and subtracted from extraterrestrial emissions. Similarly, contamination from galactic and extragalactic radio sources have to be removed. These corrections are particularly difficult at wavelengths greater than 10 centimeters where galactic emission is strong.

Measurements at wavelengths shorter than about 3 mm are even more difficult because these emissions are strongly absorbed by water vapor in Earth's atmosphere. Such observations must be performed at high, dry sites or by balloon- or rocket-borne instruments. Additional sources of noise include the telescope itself and its amplifiers and receivers. These have to be carefully measured and kept as constant as possible.

The traditional way of dealing with many of the problems with the instruments was invented by Dicke himself. Signals from the sky are compared many times each second with those from a stable reference source often immersed in liquid helium at about 4° Kelvin. Such "Dicke switching" allows very small extraterrestrial emissions to be distinguished from the random noise present in both sky and reference sources. In practice, a main and a reference receiving horn are often used. Alternately switching between the two allows imbalances in the equipment to be subtracted out.

Contributions of galactic and extragalactic radio sources are not dealt with as easily. They are generally estimated on the basis of observations at other wavelengths and the known or, more usually, assumed properties of these sources.

A few examples of pre-COBE cosmic background observations are useful to put the satellite's achievements into perspective. A collaboration of researchers from Haverford College and the universities of Milan (Italy) and California (Berkeley) led by George Smoot made precise ground-based measurements of the background spectrum at five wavelengths ranging from 3 mm to 21 cm. Their results, accurate to better than 0.1° K, are consistent with a blackbody temperature of 2.75° K.

David Johnson, David Wilkinson, and colleagues at Princeton University took a different approach. Using a sophisticated balloon-borne radiometer to minimize atmospheric contamination, they measured a blackbody temperature of 2.78 ± 0.03° K at a wavelength of 1.2 cm. This was the most sensitive measurement in the so-called Rayleigh-Jeans (long wavelength) spectral region prior to COBE's launch.

The cyanogen thermometer described earlier provides additional measurements near the peak of the blackbody curve. The method has been refined by two groups. David Meyer (Northeastern University) and Michael Jura (University of California, Los Angeles), and Philippe Crane (European Southern Observatory) and collaborators derived temperatures of 2.74 ± 0.05° K and 2.75 ± 0.2° K from observations at wavelengths of 2.64 and 1.32 mm, respectively.

At wavelengths shorter than 1 mm the brightness of the background radiation starts to drop, and conventional radio receivers are not very efficient. Measurements in this spectral region are typically made with ultrasensitive thermometers called bolometers, which are carried above the denser portions of Earth's atmosphere on balloons or rockets.

One measurement of this type caused a sensation. Andrew Lange and Paul Richards from the University of California, Berkeley, and Toshio Matsumoto and colleagues from Nagoya University found evidence for excess brightness in two wavelength bands centered at 0.7 and 0.5 mm. This distortion was so large that it amounted to 20 percent of the cosmic background's total energy.

Not surprisingly, the results of the Berkeley-Nagoya experiment stimulated a plethora of exotic theories. Some researchers sought to save the Big Bang by hypothesizing either vast numbers of dusty protogalaxies or enormous outpourings of energy into the early Universe to heat the intergalactic medium, at a rate vastly more prolific than had been envisaged by the Big Bang advocates prior to the U.S.-Japanese rocket flight.

The results were also interpreted by some skeptics as evidence against the Big-Bang theory. Rival models, such as the steady state theory, predicted such distortions. If the Universe is eternal and matter is continually created out of the vacuum to maintain a constant density of galaxies, as Fred Hoyle and others had postulated, then the microwave background was not the redshifted radiation from the cosmic fireball. Instead, Hoyle suggested, it was due to the reprocessing of far-infrared and radio photons from distant galaxies by a pervasive distribution of cold dust. Astronomers have always rejected this idea because they knew that the resulting radiation field would show gross distortions relative to that of a blackbody at 3° Kelvin.

The COBE Revolution

It was clear that COBE's instruments would be so sensitive that they could easily reveal whether or not the Berkeley-Nagoya distortion was real. Hence, it was with keen anticipation that the astronomical community greeted John Mather, principal investigator for COBE's far infrared absolute spectrophotometer (FIRAS), when he presented the first results at the January 1990 meeting of the American Astronomical Society in Arlington, Virginia. According to an eyewitness account in Physics Today, "the moment Mather placed the COBE spectrum on the overhead projector, the packed lecture hall burst into uncharacteristic sustained applause." There were no obvious deviations from a blackbody curve at all.

COBE, launched on November 18, 1989, was the culmination of a decade and a half of painstaking preparation. After only nine minutes of observing, FIRAS had obtained a spectacular result: the spectrum of the cosmic background radiation fitted a blackbody curve to a precision of 1 percent. The inferred temperature was 2.735° K.

The FIRAS is a rapid-scan interferometer cooled to a temperature of 1.5° K by liquid helium. It operates over a wavelength range from 1 cm to 0.1 mm, centered at the peak intensity of the background radiation. One antenna horn, with a 7° field of view, looks at the sky, while a similar internal horn views a temperature-controlled reference blackbody. There is also a movable external calibrator that can be periodically commanded to swing in front of the sky horn. By comparing measurements of the sky with those from the external calibrator, the blackbody spectrum can be measured more precisely than ever before.

COBE scientists anticipate that FIRAS will eventually be able to measure spectral distortions in the cosmic microwave background to an accuracy of 1 part in 1,000. It would indeed be surprising if, at this level, the long anticipated spectral deviations were not found. Protogalaxies and protoclusters must be leaving their imprint on the microwave sky if our cherished beliefs about structure and formation in the early Universe have any validity.

COBE's work is far from done. To date, the satellite's two other instruments have only reported preliminary results. Investigators responsible for the diffuse infrared background experiment (DIRBE) have issued tentative maps of the sky brightness at three wavelengths. Ultimately, the DIRBE group hopes to detect the faint glow from the first generation of stars and young galaxies.

The differential microwave radiometer (DMR) will measure variations in the intensity of the cosmic background from place to place on the sky with an ultimate precision of 1 part in 100,000. So-called inflationary cosmological models predict that variations near this level should result from the primordial density fluctuations responsible for the large-scale structures seen in the Universe today.

Perhaps the most interesting results will be those we have not anticipated.

Update: The FIRAS experiment team has analyzed the first year of COBE data. The measured cosmic microwave background is 2.726 degrees Kelvin. There are no deviations near the peak to less than three-hundredths of a percent.

An Infrared View of the Universe

It is said that the Holy Grail of infrared astronomy is the discovery of a star being born. The Infrared Astronomical Satellite (IRAS) has brought astronomers closer than ever before towards attaining it.

The satellite experiment is the result of a collaboration between British, Dutch and U.S. astronomers, and was intended to provide a systematic, unbiased and sensitive survey of the entire sky over four infrared bands, at 12, 25, 60 and 100 μm. The advantages of conducting an infrared survey above the Earth's atmosphere are twofold: first, only in space can a telescope be sufficiently cold to overcome the thermal pollution that fogs the infrared sky as seen from the ground; and second, it makes it possible to peer more deeply into the Universe. Although the satellite, launched on 23 January 1983, exhausted its cryogenic helium supply barely 10 months later, thereby terminating the survey, the accumulated data will occupy astronomers for years to come.

Preliminary results of a minisurvey over some 900 square degrees revealed nearly 9,000 infrared sources, according to Rowan-Robinson et al. Away from the plane of our Milky Way galaxy, more than half of the sources at 12 and 25 μm can be identified with bright stars. About a quarter of the sources seen at longer wavelengths are identified with known galaxies, while many others are more local objects embedded in dust clouds—perhaps newly forming stars. Within 10° of the galactic plane, there is so much activity associated with star formation that it becomes exceedingly difficult to pick out individual objects from the survey.

Study of the sources that can be identified with known objects has proved rewarding. A characteristic feature of many regions of star formation is the presence of energetic outflows, which can be detected both di-

Originally published as "An Infrared View of the Universe," Joseph Silk, *Nature*, 308, 224–225, 15 March 1984.

rectly as high-velocity molecular gas and by the presence of shock-excited gas clumps—known as Herbig-Haro objects—which have been conjectured to be the remnants of the interaction of a high-velocity wind with a dense inhomogeneous molecular cloud. Through infrared observation it is possible to probe the nature of the heavily embedded objects that drive the outflows.

Emerson et al. studied two candidate driving sources in the dark cloud L1551, which is associated with three Herbig-Haro objects and has a well-mapped bipolar flow. One source (IRS-5) had previously been identified with a low-mass protostar that drives the molecular gas outflow. The other is 0.1 pc away from IRS-5 and appears to be a dust-embedded precursor of a low-luminosity pre-main-sequence star. Star formation is occurring at more than one place in the L1551 cloud, as predicted if the parent cloud has undergone fragmentation. The situation resembles that seen in sites of massive star formation: low-mass stars may form by a similar mechanism. Another pair of Herbig-Haro objects, HH46 and HH47, appear to have been ejected from a nearby dark globule, which contains a similar dust-embedded precursor of a low-mass star.

That formation of low-mass stars is endemic to dark globules has been confirmed in more detailed studies of dark clouds. Two compact sources of radiation were discovered within the dense core of the cloud Barnard 5. One is a newly formed protostar of solar mass; the second is much cooler and may be a density fluctuation in the very earliest stage of collapse. Two other newly formed stars of solar mass, still enshrouded in dust shells, were found nearby. Thus Barnard 5, an apparently quiescent globule of dense molecular gas, has now been observed to contain as many as four protostars. The stars may have formed by the fragmentation of the dense core of Barnard 5 and are now responsible for providing enough energy input to stabilize the cloud against further collapse. The dark cloud complex Chamaeleon I was found to contain many newly formed stars. Some sources have been identified with known pre-main-sequence objects, but a considerable proportion have no optical counterpart: presumably, they are deeply embedded in the cloud. Cool sources were detected only at the longer wavelengths and appear to surround the main cloud. They seem to be associated with small globules and may represent the earlier stages of collapse. Perhaps a hundred thousand years from now, protostars of solar mass will have emerged from cocoons in the globules.

Clearly, IRAS has produced the best links, as yet, between molecular clouds and newly formed stars. But it has also allowed a fascinating insight into the nature of galaxies, because almost all the energy produced by newly formed stars is emitted from their placental gas and dust shrouds at infrared wavelengths. The infrared properties of galaxies determine the distribution, energetics and overall rate of star formation which is invaluable information for understanding galactic evolution.

In a study of a complete optical sample of 165 nearby galaxies, de Jong et al. verify that infrared emission is generally correlated with blue luminosity, which is associated with massive young stars. Predominantly spiral galaxies have been detected by IRAS; normal ellipticals in the sample were not detected.

One surprise is that the range of infrared-to-blue-luminosity varies considerably from galaxy to galaxy. While the Andromeda galaxy, for example, is a very weak infrared source, emitting only 3 per cent of its blue light in the infrared, other more active spirals emit up to five times as much infrared as blue light. The weakly emitting spirals have rather cool dust (at a color temperature of ~25K), while the more luminous infrared galaxies are often warmer (~50K), suggesting an enhanced formation rate of massive O stars surrounded by molecular clouds.

It might be predicted that a sample of galaxies selected from the infrared survey would contain the most infrared-luminous galaxies. Indeed, Soifer et al. identify nearly 100 spirals that have infrared luminosities greater than or comparable to their luminosities in the blue. The wide range in infrared luminosities is probably the result of variation in dust content from galaxy to galaxy, as well as variations in star formation rate. The very luminous infrared galaxies are only a small proportion of all galaxies. However, a surprisingly large fraction of the infrared-selected sample are found to have neighboring galaxies nearby. Theory has suggested that the tidal interaction of a galaxy with close neighbors can trigger outbursts of star formation, and the IRAS data provide tantalizing evidence for this conjecture. Moreover Young et al. found that many of the infrared sources detected in the Hercules cluster of galaxies are spiral galaxies. The most luminous galaxy has a warped disc, and has been catalogued as an interacting galaxy. Its enhanced infrared luminosity is probably a result of a burst of star formation triggered by the interaction: tidal forces would have compressed molecular cloud complexes and initiated collapse.

Since some galaxies can be such spectacular emitters in the far infrared, infrared emission might reveal the presence of galaxies where optical observations are inconclusive. This seems to be the case for quasars. In a comparison of the infrared data on radio-loud and radio-quiet quasars, Neugebauer et al. find the infrared continua of the radio-loud quasars are consistent with the spectrum expected from synchrotron emission by relativistic electrons, as extrapolated from the radio-frequency region. By contrast, the two radio-quiet quasars observed reveal an excess over the power-law component at 100 μm, suggesting a contribution from an underlying spiral galaxy that is vigorously forming stars and is bright at 100 μm. Indeed there is evidence at 1–2 μm that the infrared emission around both radio-quiet quasars is extended, consistent with the presence of a surrounding galaxy. The infrared 100 μm luminosity for the underlying galaxy is similar to that of active spiral galaxies.

One of the most spectacular results from IRAS is the discovery by Auman et al. of an infrared excess, extending to 20 arc s at 60 μm, around the bright star Vega. The most likely origin of the infrared emission is thermal radiation from solid grains, at least 1 mm in radius—smaller grains would have spiralled into Vega by the Poynting-Robertson effect—and heated by light from Vega to about 85K. The grains must lie in a shell or ring at a distance of about 85 AU from Vega and in orbit around the star. These particles cannot have been present since the time of formation of Vega, some 1–2 $\times 10^8$ years ago, since they are much larger than typical interstellar grains (which are at most 0.5 μm in size). Clearly, they have grown considerably since Vega formed. Since our Solar System is about 4.5 $\times 10^9$ years old, it is tempting to infer that the shell around Vega represents some intermediate stage in the formation of a planetary system.

Another major surprise from IRAS is the discovery of interstellar cirrus. Low et al. report extended patches of far-infrared emission, seen predominantly at 60 μm and 100 μm, both high above the galactic plane and in the ecliptic plane. Three components have been identified: one associated with neutral hydrogen and dust concentrations at very high galactic latitudes; another in the ecliptic, associated with the interplanetary medium and apparently running continuously around the Solar System; and a third—the coldest—which correlates poorly with any known structure in the Galaxy or Solar System. Degree-size patches of 100 μm emission are seen and could represent either dust clouds in the outer Solar System, or a novel component of the interstellar medium. IRAS observations, taken six months apart, should be capable of resolving this enigma.

Finally, the IRAS minisurvey revealed some very intriguing unidentified objects that may be of a fundamentally new type. Houck et al. identify nine point sources that are bright at 60 μm but have no obvious identifiable optical counterpart brighter than 18.5 mag. The intrinsic infrared luminosity exceeds that in the blue band by at least an order of magnitude. Of course, these sources may not be extragalactic or even outside the Solar System. Nevertheless, their discovery is one of the potentially most important accomplishments of IRAS. Radio surveys led to the discovery of quasars and pulsars; X-ray surveys came up with X-ray binaries and bursters; now the race will begin to explain the unidentified sources in the first all-sky infrared survey.

During the next decade, both the European Space Agency and the National Aeronautics and Space Administration have plans for a sequel to IRAS. Both the European (ISO for Infrared Space Observatory) and the American (SIRTF for Shuttle Infrared Telescope Facility) experiments are designed to be operated as true observatories for infrared astronomy, carrying a variety of focal-plane instruments. Cryogenically cooled 1-m class space telescopes will be able to study the infrared sources discovered by IRAS at more than 100 times the sensitivity and spectral resolution and provide detailed pictures of even the faintest IRAS sources. This quantum leap in infrared capability will do for the infrared astronomers what the Einstein Observatory did for X-ray astronomers. Only then will we see more than the tantalizing glimpse of the infrared sky that IRAS has already provided.

Dark Matter Detected?

Yet another Holy Grail of cosmology may have been detected in October, 1993. This month saw the announcement that a candidate for the dark matter in the halo of our galaxy has finally been identified.[1,2] It is perhaps a fitting finale to this collection to end by recognizing that perhaps dark matter consists of the most mundane stuff: stellar mass objects. There may well be, and indeed must be, more dark matter than is seen in galaxy halos. Whether it has the same form as halo dark matter or is of a more exotic nature is unknown.

The rotation speed of stars and gas clouds around the Milky Way unambiguously specifies the dark matter content of the halo of our galaxy. Near the sun, this amounts to 0.01 $M_{\odot}$ per cubic parsec, or about one-tenth that of local luminous matter. However, the luminous matter is concentrated within a disk of thickness several hundred parsecs, whereas the dark matter occupies a quasi-spherical region that has a core radius of about 10 kiloparsecs, and extends to at least 30 kiloparsecs. The luminous mass of the Milky Way is 10^{11} $M_{\odot}$, and the dark matter contributes a factor of 10 more mass that mostly is in the outer halo. Ninety percent of our galaxy is dark, and elucidating its nature has provided one of the decade's greatest challenges to astronomers.

Theory suggests two distinct alternatives for the dark matter. One option favored by the particle astrophysicists is particle dark matter, in the form of weakly interacting relic particles left over from the Big Bang. The very early Universe provided an unexcelled particle accelerator, one that outperforms the Superconducting Super-Collider, the Large Hadron Collider or any conceivable future machine, by many orders of magnitude in achievable energy. The cosmos once contained matter at a characteristic energy of 10^{16} TeV: the LHC will achieve energies of a few TeV. Of course we cannot do controlled experiments, but we can study any surviving relics from the one great experiment in the sky. Such extreme energies did not last long, only 10^{-43} s at 10^{16} TeV, and 10^{-12} s at 1 TeV as the Universe expanded and cooled. These energies sufficed to create particles, along with antiparticles, of all possible varieties. Every known, and unknown, denizen of the zoo of elementary particles existed, however briefly, in the very early Universe. As the temperature fell, the new particles, almost invariably short-lived, were no longer created. Those few remaining annihilated until the density dropped sufficiently that a few stragglers remained. Protons are survivors, along with a very few antiprotons. Indeed, the only reason that protons dominate over antiprotons today is that an asymmetry was imprinted early in the history of the Universe that created an excess of protons over antiprotons. The relic antiprotons that survived amount to only about one antiproton for every 10^9 protons. However any stable particles that interact more weakly than protons would have survived in greater numbers.

Supersymmetry (or SUSY) doubles the number of known particles. For every elementary particle of spin 1 (or 1/2), the theory of supersymmetry postulates a particle of spin 1/2 (or 1). This leads to the prediction of the photino, partner to the photon, the sneutrino, partner to the neutrino, the wino, partner to the W-boson, the higgsino, partner to the elusive higgs boson, the zino, partner to the Z-boson, and so on. Theory predicts the existence of these particles, but none have yet been discovered. One reason for believing that supersymmetry is a valid theory is that recent measurements at LEP of the energy dependence of the weak and strong nuclear interactions suggest a convergence to a unique energy scale only in supersymmetric models.[3] This energy scale, although admittedly an extrapolation from measured energies by many orders of magnitude, corresponds to that of the grand unification of the electromagnetic, weak and strong interactions at about 10^{16} GeV. The energy scale at which supersymmetry was broken must be greater than the electro-weak unification scale of 200 GeV, or signatures would already have been detected in accelerator experiments. Perhaps the SSC will find evidence for supersymmetry. Cosmology, however, does provide a laboratory where supersymmetry once reigned: the Universe is not supersymmetric today, but once it must have been.

Almost all of the supersymmetric partner particles decayed once particle energies fell below the supersymmetric threshold energy. However the lightest supersymmetric partner, generally believed to be the photino, must be stable: all heavier superparticles decayed, leaving an ineradicable trace abundance of photinos. Photinos annihilate, since they are their own antiparticles, but as the expansion continues and the density drops, a relic abundance of photinos survives. The photino is a massive, weakly interacting particle. Supersymmetric models specify its cross-section for interaction with ordinary matter to be weak, and similar to that of neutrinos if the supersymmetric particle mass is on the order of that of the electroweak unification scale. Cosmology specifies the relic photino abundance, once its cross-section for survival against annihilation is specified. Remarkably, for a typical weak cross-section, the predicted abundance is within an order of magnitude of the critical density for closure of the Universe. Accelerator measurements and cosmological bounds jointly constrain its mass to be at least 30 GeV.[4] While cosmology does not guarantee the photino's existence, it does predict its abundance, given that supersymmetry is valid.

One therefore has in the photino, or in its generic form, the neutralino, a dark matter candidate. Dark halos may therefore consist of such particles, and the acronym WIMP, for weakly interacting massive particle, has been coined to describe the SUSY stable relics. Various experiments are underway to detect WIMPs, none of which have hitherto reported any success. WIMPs annihilate today, albeit at a very slow rate, but the annihilation products may be detectable if the WIMP densities are sufficiently large. One site of interest is the galactic halo, where WIMP annihilations produce high energy (GeV) gamma rays, that are potentially detectable as a diffuse gamma ray glow from the halo by space satellite experiments such as the Gamma Ray Observatory. A second site is the interior of the sun: as the sun orbits the Galaxy, it sweeps up WIMPs that accumulate in the core of the sun, where they annihilate. The high energy neutrinos produced in the annihilations are far more energetic than the neutrinos produced by thermonuclear fusion in the sun, and are another WIMP signature that is visible on the Earth in deep underground detectors. These detectors, running at the Gran Sasso Underground Laboratory in Italy, under a mile of rock, and in the Kamiokande mine in Japan, are searching for high energy neutrinos from cosmic sources: so far, no WIMP events from the sun have been seen.

There are also several experiments underway around the world, in France, Germany, Italy, the U.K., and the U.S., to directly detect the WIMPs. These experiments utilize the nuclear recoil induced by a WIMP hitting a heavy element nucleus, and study the resulting phonon and ionization signals in a cryogenically cooled system. Again, given the WIMP abundance near the sun, a detectable event rate is achievable with experiments presently under design. For all three types of experiments, if the WIMPs have masses below about 100 proton masses, there should be enough WIMPs, and enough annihilation events, to eventually give a signal if WIMPs are indeed the halo dark matter.

The alternative dark matter candidate to a WIMP is some form of suitably dim star or star-like object. These are generically given the acronym MACHO, for massive astrophysical compact halo objects. Plausible astrophysical MACHO candidates, objects that are actually known to exist, unlike WIMPs, include brown dwarfs, or sub-stellar mass objects, white dwarfs, neutron stars, and black holes. While all of these are known, or strongly presumed to exist, one has no prediction of their likely abundance in the halo. The history of star formation in the early galaxy is too uncertain to be a reliable guide.

One can appeal to indirect arguments: for example, black holes weighing more than about 1000 $M_\odot$ have been eliminated as a possible candidate on the grounds that they would tidally disrupt the observed halo globular star clusters.[5] One can argue that neutron stars or stellar-mass black holes are unlikely candidates, because of the implied vast number of stellar explosions, supernovae, and associated metal-production and ejection that gave rise to them.[6] The best MACHO bets are almost certainly brown dwarfs, which span the range from giant planets ($\sim 10^{-3}$ $M_\odot$) to objects at the threshold of hydrogen burning (0.08 $M_\odot$, for solar metallicity, to about 0.1 $M_\odot$ for a primordial abundance mixture), and white dwarfs, expected to span the mass range 0.4–1.4 $M_\odot$. However, one cannot exclude the more exotic objects, nor even planet-mass objects, other than by the argument that these latter objects, in so far as is known by observing the Solar System, provide an exceedingly small fraction of dark relative to luminous matter.

By contrast, if known mass fraction were one's criterion of choice, white dwarfs are the clear winner, since these objects, which are dark and almost invisible after a few billion years, constitute up to 20 percent of the mass of the oldest known stellar systems in the halo, namely globular star clusters. One would have to presume that early in the history of the galaxy, star formation somehow favored the more massive stars in the range 2–8 $M_\odot$ that generated white dwarfs. This is not an easily digestible assumption, although it is consistent with all astrophysical constraints.[7] Indeed, a halo of white dwarfs provides a reservoir of objects for occasionally forming neutron stars by accretion-induced collapse in binaries: these may even have been seen as high velocity radio pulsars and gamma ray bursters.[8] A far more innocuous choice would be brown dwarfs, for which however the evidence of existence even in the Galactic disk remains sparse, despite intensive searches. Nevertheless, star formation theory makes their existence plausible, and studies of the local initial stellar mass function suggest that as much as 10 percent of the mass in stars could be in brown dwarfs. The ultimate arbiter has of course to be experimental detection.

Results from two experiments that find strong evidence for the existence of MACHOs were reported[1,2] in October 1993. The technique used is gravitational microlensing.[9] If a MACHO passes very close to the line-of-sight from Earth to a distant star, the gravity of the otherwise invisible MACHO causes bending of the starlight and acts as a lens. For halo MACHOs, the star splits into multiple images that are separated by a milliarcsecond, far too small to observe. However, the background star temporarily brightens as the MACHO moves across the line-of-sight in the course of its orbit around the Milky Way halo.

There are two major difficulties with this experiment. First, the microlensing events are very rare: only about one background star in two million will be microlensed at a given time. Secondly, many stars are intrinsically variable. However the microlensing event has some unique signatures that help distinguish it from a variable star. It should be symmetrical in time, achromatic, and should occur only once for a given star.

To overcome the low probability, two experiments were designed to monitor millions of stars in the Large Magellanic Cloud. One group, a collaboration of French astrophysicists at Saclay, Orsay, Marseille, and Paris, led by Michel Spiro, utilized a total of more than 300 ESO Schmidt plates taken of the LMC over a 3 year period with red or blue filters. An analysis of 2.5 million stars revealed two events that displayed the characteristic microlensing signatures, with an event duration of about 50 days. The second group is a U.S.–Australian collaboration led by Charles Alcock that utilizes the 50-inch telescope at Mt. Stromlo, refurbished and dedicated to the MACHO search, in conjunction with the world's largest CCD camera built for astronomical use by Christopher Stubbs at the Center for Particle Astrophysics, University of California at Berkeley. The astronomers, from LLNL, MSSSO, Berkeley, Santa Barbara, San Diego, and Ann Arbor, report data on 1.8 million stars, observing each star about 250 times. After analyzing 15 percent of the data taken during the first observing season, they found one "gold-plated" microlensing event with a duration of 34 days. The peak increase in brightness was a factor of 7.

The duration of the microlensing event directly measures the mass of the MACHO, with some uncertainty because of the unknown transverse velocity of the MACHO across the line-of-sight. The duration of the event is simply the time for the MACHO to cross the Einstein ring diameter. The Einstein ring radius is a few astronomical units, being approximately equal to the geometric mean of the Schwarzschild radius of the MACHO and the distance to the MACHO, for a MACHO half-way to the LMC, distance 55 kpc. The event durations suggest a typical mass around 0.1 $M_\odot$, but with at least a factor of 3 uncertainty.

Much more data remains to be analyzed by the two groups. The MA-

CHO interpretation, if correct, should result in more events that are distributed according to the expected distribution both of amplifications and of the properties of the background stars. In the meantime, one can speculate about the implications. The three events detected are to within a factor of 2, what the MACHO model of dark halo matter predicts. This certainly presents the strongest evidence to date of dark matter detection. Unless there are perverse types of rare variable stars, MACHOs constitute a significant fraction of the dark halo. Suppose, however, that further events are not forthcoming in significant numbers. A chastening thought is that the MACHO experiments have signally failed to report events of duration shorter than 30 days in their initial data analysis. If this trend continues to reflect a paucity of MACHO objects below 0.1 $M_\odot$, admittedly a premature deduction in view of the uncertain sensitivity of the experiments according to presently available information, the most natural interpretation of the microlensing is in terms of M-dwarfs, hydrogen-burning stars. These would not be halo dark matter, but rather constituents of the thick disk of the galaxy. Indeed, a third microlensing experiment by a Polish–U.S. collaboration at the Las Campanas Observatory, looking towards the bulge of our Galaxy, has reported a microlensing event that, if real, is produced by a low mass main-sequence disk star.

The MACHO interpretation, if confirmed, will be welcomed by astrophysicists. The interpretation of light element abundances in terms of primordial nucleosynthesis in the early Universe already requires a substantial amount of baryonic dark matter, equivalent to about 3 percent of the initial density.[10] Coincidentally, this is also approximately the total mass in dark halos, thereby raising the possibility that one could be finally detecting the predicted baryonic dark matter. Such a conclusion, if the Universe is at critical density as predicted by inflationary cosmology, implies that essentially all of the baryons formed dark halos and galaxies, leaving behind the non-baryonic matter to permeate intergalactic space. How much of the non-baryonic matter was entrapped within halos is an interesting question, of crucial importance for the experimenters undertaking direct detection searches for WIMPs.

Suppose either that the Local Group is predominantly cold dark matter, or else imagine the halo collapsing spherically and starting from the mean density of the Universe about 10 billion years ago. One arrives at a cold dark matter overdensity in the region where our galaxy halo is forming. As baryons collapse to form the halo, adiabatic enhancement due to the self-gravity of the baryonic halo must locally enhance the cold dark matter density. A crude minimal estimate is that the non-baryonic dark matter density in the halo should be at least one percent that of the halo dark matter density. This means that the direct detection experiments may have to try, in a worst case scenario, 100 times harder than previously anticipated to search for a WIMP signature.

If one indeed has detected baryonic dark matter, then on the grounds of simplicity, one might alternatively try to argue that there is no longer any compelling need for WIMPs. In this case, it is difficult to arrange to have a critical cosmological density in MACHOs. Problems arise in such a high baryon density Universe with producing excessive fluctuations[11] and spectral distortions[12] in the cosmic microwave background radiation, as well as with having to reject the primordial nucleosynthesis predictions. A more likely scenario would be a universe containing about one-tenth of the initial density, precisely what is observed by astronomers who study the nearby Universe over scales of up to a few megaparsecs. Such a model, a baryonic open universe, meets most of the large-scale structure constraints from galaxy clustering, provided that one adopts a suitable choice for the spectrum of primordial density fluctuations.[13] Inflation no longer provides the initial conditions if the Universe is open, so that this model is phenomenologically flexible.

Whichever choice is preferred, inflationary cold dark matter with baryonic dark halos, or an open, baryonic dark matter-dominated universe, the theory of galaxy formation now takes on a new dimension. Most baryons form MACHOs in the halo, and the remaining baryons form stars in the inner spheroid and disk. This will surely lead to novel predictions for galaxy evolution and the search for protogalaxies. An early luminous phase, when the halos formed, might well provide a detectable signature of baryonic dark halos in the early Universe.

1. Alcock, C. et al., Nature, 365, 621 (1993).

2. Aubourg E. et al., Nature, 365, 621 (1993).

3. Amaldi, U., de Boer, W., & Furstenau, H. Phys. Lett B260, 447 (1991).

4. Olive, K.A. & Srednicki, M. Nucl. Phys. B355, 208 (1991).

5. Moore, B. Astrophys. J. 413, L93 (1993).

6. Ryu, D., Olive, K.A., & Silk, J. Astrophys. J. 353, 81 (1990).

7. Silk, J. Science 251, 537 (1991).

8. Eichler, D., Silk, J. Science 257, 937 (1992).

9. Paczynski, B. Astrophys. J. 304,1 (1986).

10. Walker, T.P. et al. Astrophys. J. 376, 51 (1991).

11. Hu, W., Scott, D., & Silk, J. Astrophys. J., in press (1993).

12. Tegmark, M. & Silk, J. Astrophys. J., in press (1993).

13. Cen, R.Y., Ostriker, J.P., & Peebles, P.J.E. Astrophys. J., 415, 423 (1993).

Acknowledgments

Several of the articles in this book were written with coauthors to whom I am profoundly grateful for sharing inspiration and time in the mutual goal of reaching a broader audience—one that is not accustomed to browsing in the pages of the *Astrophysical Journal* or *Physical Review*. My collaborators include Beatriz Barbuy, John Barrow, Jim Bartlett, Roger Cayrel, Gus Evrard, Roman Juszkiewicz, Cedric Lacey, Martin Rees, Alex Szalay, David Weinberg, and especially someone who was a great inspiration to me, the late Yakov Zel'dovich. I am also indebted to many of my colleagues for the innumerable coffee-time and tea-time discussions that helped to enrich the articles in this collection.

Index

About the Author

Joseph Silk is among those rare scholars who not only contribute handsomely to science, but also write about it with a command that captures the admiration of his colleagues and the delight of the public. A member of the International Astronomical Union, the American Astronomical Society, the American Physical Society, and a Fellow of the Royal Astronomical Society, Professor Silk teaches at the University of California at Berkeley, where he does research in theoretical astrophysics and works at the interface of particle physics and cosmology.

In addition to his position as Professor of Astronomy and Physics at Berkeley, Silk has held teaching and research posts at Canberra, Paris, Princeton, and Cambridge. He has delivered more than a hundred invited lectures at scientific meetings and has been the co-chair of four international conferences. Since 1992, he has been a member of the Aspen Center for Physics.

Alone, or with others, he has written three widely acclaimed books on astronomy and cosmology—two fascinating accounts for general readers, *The Big Bang* (Freeman) and *The Left Hand of Creation* (Basic). He is also the author of more than 200 scientific papers, 70 review articles, and 50 popular articles.

Born in London, Silk received degrees in mathematics from Cambridge and his doctorate in astronomy from Harvard. For his impressive contributions in physics and astronomy, Dr. Silk has been awarded numerous honors, among them Harvard's Bowdoin and Bok prizes and Sloan and Guggenheim fellowships. He has been Leon Lecturer at the University of Pennsylvania and was the Hooker Distinguished Visiting Professor at McMaster University. In 1987, he was made a Fellow of the American Association for the Advancement of Science.

DISCARDED